MW01362129

# WAVE CATCHER

# WAVE CATCHER

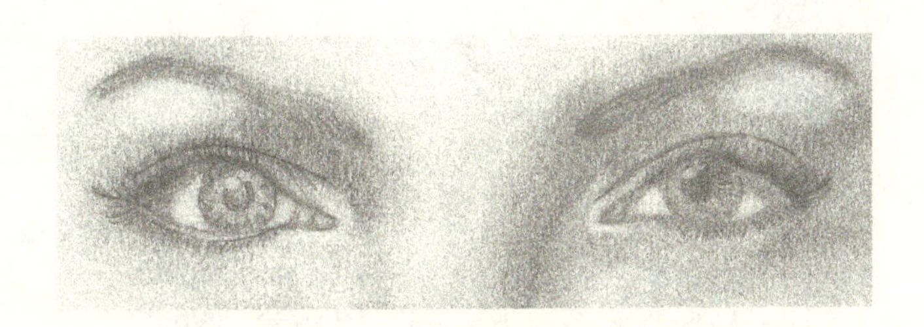

It has alluring curves
It's whispered into
It awakens you

A discovery by a world-renowned neurologist
that will uncover a treasure inside of you.

DR. A.C. KALFUS

| Library of Congress Control Number: | | 2016908882 |
| --- | --- | --- |
| ISBN: | Hardcover | 978-1-5245-0608-7 |
| | Softcover | 978-1-5245-0607-0 |
| | eBook | 978-1-5245-0606-3 |



Print information available on the last page.

Rev. date: 06/30/2016

**To order additional copies of this book, contact:**
Xlibris
1-888-795-4274
www.Xlibris.com
Orders@Xlibris.com
735649

# CONTENTS


1. Wave Catcher ........ 1
    - Examples of auricular therapy ........ 3
    - A healing language ........ 5
    - Frequencies affect our health ........ 6
    - The largest and most sophisticated pharmacy in the world ........ 7
    - Our skin and our health ........ 8
2. Words Matter ........ 10
    - Research that shows how words and thoughts affect your brain ........ 12
    - Just breathe ........ 15
    - Diaphragmatic breathing ........ 16
3. Ears Do More than Capture Sound Waves ........ 18
    - Different uses of ears ........ 19
4. A Key to Health and Aging ........ 21
5. Chronic Disease May Be a Food-Borne Illness ........ 25
    - How do we become allergic to food? ........ 29
    - The first four to six months of life are unique ........ 30
6. Everything Seems to Be Turned Upside Down ........ 35
    - An upside down map of our body is found on the ear ........ 36
    - We see the world upside down ........ 37
7. The History of Auricular Medicine ........ 40
    - The oldest medical documents known to mankind ........ 44
    - The father of Greek medicine and the Roman Empire ........ 46
    - The loss of medical knowledge ........ 47
    - The beginning of hospitality ........ 48
    - The Renaissance ........ 49
    - The first publication of acupuncture in the United States ........ 50
    - The discoveries of Paul Nogier ........ 52
    - From Nogier to today ........ 53
8. A Truth Spoken Before Its Time Is Dangerous ........ 55
    - The parable of the blind men and an elephant ........ 56

9. What Makes Chronic Pain Unique? ..... 57
10. The Virtue of a Touch ..... 59
    What we can see when we open our eyes to our ears ..... 60
    Pain is both a personal and public health challenge ..... 64
    The new normal ..... 65
11. Common Myths about Auricular Therapy ..... 66
12. Ear Points Are Strange and Beautiful to Well Opened Eyes ..... 70
    Two types of ear points ..... 72
    A Reflex point ..... 73
    Our skin receives and interprets light ..... 76
    Our ear is uniquely wired to the entire body ..... 78
    A simple way to elicit ear pain ..... 82
    Electrical detection of an ear point ..... 84
    Four distinct characteristics of each active ear point ..... 87
    Our ear helps to regulate our body temperature ..... 87
    The ear and its immune system guardians ..... 89
    Treating the ear using electrical stimulation ..... 89
13. In the Beginning ..... 91
    The light of life ..... 92
    The seven frequencies of light that connect all animals on earth ..... 93
    The story of cells and their love of light ..... 93
    Our skin is like that of a Chameleon ..... 97
    In the complexity of a problem, is the simplicity of the solution ..... 99
    The frequencies are related mathematically ..... 100
    A frequency that sets people apart ..... 109
    Precise diagnosis of left and right brain communication disorders ..... 111
    Humanity has a dominant side ..... 112
    Less is better ..... 114
    Light and depression ..... 115
14. Obstacles in the Treatment of Chronic Pain ..... 117
    First rib syndrome ..... 118
    How to diagnose a First Rib syndrome ..... 122
    Dental focus ..... 123

A painful message can remain long after its source is gone ...126
How to diagnose a dental focus ...126
The Nogier Pulse ...129
15. Look for What Is and Not for What Should Be ...131
The Nerve that lowers our stress level ...133
The number 40 ...133
The origins of electrically charged treatment points on the ear ...134
16. It's Better to Change an Opinion Than to Persist in a Wrong One ...135
17. Even Though Ear Points Move, We Will Find Them ...138
Everything's in motion ...140
What does a change in the strength of our pulse mean? ...141
How Dr. Nogier discovered the pulse reflex ...142
Obsession ...144
The rabbit test ...145
A simple and comfortable way to test for allergies ...147
18. Our Ear Is Intimately Linked to Every Cell in Our Body ...148
19. The Science behind Auricular Therapy ...150
20. Physiological Basis for Auricular Therapy ...153
21. What Conditions Has Auricular Therapy Been Proven to Benefit? ...156
22. Is Chronic Pain Due to An Alteration of Natural Healing Neural Circuits? ...158
23. How Auricular Therapy Affects the Neurophysiology of Pain ...159
24. All Living Cells Emit Light ...163
25. The Hub of Life Is Water and Light ...167
The Liquid Crystalline Form of Water ...169
26. How Ear Acupuncture Relieves Pain ...180
27. Beauty Is More Than Skin Deep ...183
28. Physiological Features Unique to Auricular Reflex Points ...185
29. Identifying Neural Correlates with Auricular Therapy Using Brain Imaging ...195
30. Identifying Endorphin and Neurochemical Correlates Following Auricular Stimulation ...199

31. It Can go Either Way ..... 201
32. Things Appear Simple from a Distance ..... 205
33. Embryological Derivatives on the Ear ..... 208
34. Two-Way Reflexes on the Ear ..... 210
35. Diagnostic Value of Visual Changes on the Ear ..... 212
36. Illustrations ..... 215

# Preface

Pain—it's a stubbed toe. It's a headache. It's a backache, a stomachache or a shoulder ache that won't go away. We all deal with it one way or another. But isn't it supposed to protect us from something that's going to harm us? Pain can be a good thing. It's meant to tell us that something is wrong and to teach us which behaviors are safe and which ones aren't. It is supposed to turn our wounds into wisdom. But what happens when it sets up residency in our life and becomes chronic? The physical and emotional pain that was supposed to help us can now become harmful and lead to anxiety, depression, insomnia, and drug dependency, to name a few.

Knowing that so many suffer from acute, chronic, and degenerative pain, I applied my skills as a Doctor of Dental Surgery and expanded my knowledge and expertise in pain management while serving as a Lieutenant Commander in the United States Navy. Having worked in pain clinics throughout the country treating Marines who suffered from headaches, TMJ dysfunction, and neck and shoulder pain, I began to notice an unusual trend. And before long, I was captivated. Maybe it was the unknown that I found so attractive, or maybe it was just something that I saw, which was so different from anything that I had ever seen before. Whatever the reason, finding out about this little known procedure began to monopolize so much of my time that it was hard to tell if it was seeking me or if I was seeking it. As I learned more about it, little details and preferences began to build and the most hidden secrets, which were the most difficult to know, became the most fulfilling and the most alluring. The more puzzles of information that I had to put together only meant that there would be more individual pieces to fit. At times, things began to move so fast that what may have appeared to be stillness in my fascination were actually discoveries that kept me transfixed. So what was it that captivated me? What was so fascinating? What was it that drew me in like gravity? It was an unexpected connection between people who were suffering in pain and their profound relief from pain anywhere in the body when a particular type of treatment was used. This treatment didn't require any prescription medications, and its benefits were often immediate and long lasting. In addition to feeling better, those receiving this therapy were also sleeping better and feeling less fatigued. And it was safe, comfortable, and there were no adverse side effects. Sadly, there were only a few doctors who knew about it, and those who did only used it sparingly.

In everyday life, we experience ebbs and flows that are similar to the high and low tides that come in or leave the coast. Although we can't control them, we can control how we respond. Just like ships that need to seize the high tides to enter or leave a port, it's up to us to recognize whether something that happens in our life is a blessing or an obstacle; it's up to us to seize opportunities whenever they arise before we lose them. The difference between success and failure is sometimes as simple as having good timing and knowing when the "tide" is highest. And the timing here couldn't have been better.

Just weeks before he was set to retire, a pain management specialist in the Navy referred me to a well-known oncologist who was giving a conference on the very subject that had intrigued me so much. Following the conference, I flew to the East Coast for additional training at Walter Reed Medical Center. I was then introduced to the internationally acclaimed author, lecturer, and pioneering researcher Dr. Terry Oleson, who in turn inspired me to train at Johns Hopkins University with the world-renowned French Neurologist Dr. Raphael Nogier, M.D. Dr. Nogier is the top expert in this field, the field known as Auricular Medicine. Unlike any other field in healthcare, Auricular Medicine focuses on an area that provides unprecedented access to our nervous system and our immune system to help alleviate pain and promote healing naturally. So finally, I was able to put the pieces of this fascinating puzzle together, and it's my hope to share Dr. Nogier's discoveries and the available treatments with you.

L. C. Kalfus

This unique and uncensored book will introduce you to an elegant and wonderfully effective therapy. It's intended for anyone who wants to live a more pain-free life, accelerate their natural healing process, and restore and maintain their health. "Wave Catcher" is the fruit of years of work and the result of training directly with the world's top experts in the field of auricular medicine. Up until now, few people in the United States have ever seen or let alone experienced this remarkable treatment. It's my hope to change this and to share with you a very safe, fast, and effective way for relieving the pain and functional problems associated with so many medical complaints. It's also my hope that this treatment becomes available to all of us and is used as an effective option or adjunct to drugs, surgery, and exercise.

*What if* you discovered a switch that could help you to turn on and turn off your nervous system?

*What if,* in addition to the food, water, and air that you use to maintain your health, you could also use light?

*What if* you could tap into an intimate connection that your skin has with your immune and nervous systems, and every single cell in your body could benefit?

*What if* you discovered a way to live a more pain free life that could accelerate the healing process without the use of drugs and without any adverse side effects?

*What if* the dramatic rise in chronic diseases such as heart disease, respiratory disease, and autoimmune disease, is related to the food you eat and other choices you make?

If you want to know the answers to any of these *what ifs*, then we need to talk, and through these pages, we will. First, I have some bad news and some good news. The bad news is that some of the choices you are making are contributing to your circumstances. The good news is that some of the choices you are making are contributing to your circumstances. And now is the time to make choices in your life that can help you to relieve your pain and live a more joyful life. Through "Wave Catcher", I will help you do just that.

# Acknowledgments

The writing of this book has been a wonderfully transformative yet sometimes difficult experience for me personally. It has been an undertaking that I couldn't have completed without the help and support of a number of important people in my life.

Thanks first to my family. They have been inspirational and a guiding force in my life. It's their belief and support that gave me the courage to be who I truly am instead of what others may have wanted me to be.

Thanks to my parents, especially my dad, for instilling in me the love of learning.

I especially would like to thank Raphael Nogier, M.D., for helping me to understand his thinking about auricular medicine and for sharing with me his insight and expertise on frequency therapy. My thanks also goes to Terry Oleson, Ph.D., who has helped me to understand the rich history of this specialty, the science that validates this therapy, and how best to utilize the various techniques from around the world. Terry, without your help, none of this would have been possible. My thanks to Richard C. Niemtzow, M.D., Ph.D., MPH., for sharing with me his clinical expertise and pearls of wisdom on advanced Battlefield Acupuncture, which he developed to help those who serve in our military. To Jim Shores, Ph.D., whose knowledge and expertise in bipolar electrodes, electrotherapy, and anatomy is invaluable. To John Howard, L.Ac, Dipl, Ac., for sharing with me his tremendous clinical knowledge and insight in pain management as well as providing a mathematical relevance to auricular therapy.

Last but not least, thanks to Captain Mark Roback and Captain Andy Branch of the U.S. Navy whose support and encouragement have guided me in my pain management vocation. Mark, your unyielding support was invaluable, and you alone gave me the opportunities to work and study with the world's top experts in this field. Andy, not only do I cherish the times we worked together and your Tennessee accent, but the phone calls you made on my behalf to get me in touch with the right people at the right time were priceless.

It's important to note that auricular therapy is an ever-changing science that undergoes continual development due to ongoing research and clinical experience. The author has made every effort to ensure that the information presented in Wave Catcher is both accurate and at the current state of knowledge at the time of production of this book.

May this book help in the relief of those who suffer from pain and disease so that they can once again enjoy the amazing gift of life.

# Chapter 1

## Wave Catcher

It has alluring curves; it's teeming with alarms; it awakens you; it's whispered into; it gives you a sense of balance; it's adorned; and you lend it to others. Have you guessed yet? One more hint - it's a wave catcher. It's your ear! Your brilliant and intricate ear. And far from being a left over remnant of evolution that has no functional significance, as Charles Darwin concluded in 1907, your outer ear does much more than simply catch sound waves or act as an amplifier. It's true purpose is often forsaken and goes largely unnoticed, like the stars that fade at dawn. Your ear is endowed with treasure maps that will uncover something far more valuable than money, jewels, or gold; the treasure, you see, is your health. Yet, all too often, the hardest thing to explain is that which is glaringly evident, which nobody by chance ever observes. A quotation attributed to the French physiologist Claude Bernard has always inspired me:

> It has often been said, that in order to discover things, one must be ignorant. It is better to know nothing than to have certain fixed ideas in one's mind, which are based on theories which one constantly tries to confirm. A discovery is usually an unexpected connection, which is not included in some theory. A discovery is rarely logical and often goes against the conceptions then in fashion.

The French have a beautiful proverb, *Chacun voit midi a sa porte.* The literal translation goes, "Everyone sees noon at his doorstep." It means that every individual is occupied, first and foremost, with his or her own personal interests, and sees their subjective opinions as objective truths. But what happens when something is staring you in the face and you can see it with your own eyes and feel it within your heart; do you tell yourself that only a fool doesn't believe it to be true? Or do you continue to look away and hold on to an unquestioned belief without critical examination?

After years of seeing the limitations, the lack of good outcomes and the side effects associated with our current methods of pain management, I asked myself if there could be a better way to help and heal; one with greater predictability and without the side effects and dependency of drugs. And there it was, glaringly evident and unexpected, and definitely not in fashion; at least not yet. Shakespeare once said, "It's not enough to speak, but to speak true," so I have written this book to share with you some alluring truths that will help to relieve the pain and functional problems that are associated with so many medical complaints.

Medicine has always suspected some relationship between the ear and the rest of the body, and in 1967, the father of auricular medicine, Dr. Paul Nogier, M.D., of Lyon, France, had proven it. Yet nearly a half-century later, the excitement that stirred the medical circles throughout Europe is still not being embraced here in America, and this is with an unprecedented volume of research validating its benefits.

*Wave Catcher* will introduce you to a type of treatment that requires no drugs, no medicines, and no supplements and will accurately detect and treat ailments throughout your entire body using Food and Drug Administration (FDA) approved diagnostics and advanced treatment methods. Although the research and technology behind it is sophisticated, its application is simple. It uses either precise frequencies of pulsed light, very low-voltage current at millionths of an amp, or tiny needles on the ear alone, which is recognized by the U.S. National Institutes of Health for its evidence-based value in conventional medicine as well as by the World Health Organization.

Auricular therapy helps to revive and accelerate our natural healing process. Our skin, and in particular the skin around our ear, shares valuable information about our health, and this can be used to help us. In my experience, a reduction of pain is often followed by a renewed sense of hope and with hope, anything is possible. Regrettably, the solutions to many of our problems are often right in front of us, yet we fail to recognize them. But when we do *lend an ear* to what our body is saying, the obstacles to our health can be cleared away and "voila," we can begin to feel better again!

## Examples of auricular therapy

To better understand auricular medicine and how it works, I thought it would be helpful to share a few stories with you. Throughout this book, I'll share examples of actual patient treatments and outcomes so that you can see just how safe, simple, and effective this therapy is. Just as it has helped countless others, it can help you too.

In the first story, a thirty-four-year-old woman was referred to me after seeing a number of specialists. She was no longer able to work and on full disability. Her story began a few years ago when she was doing something she loved; she loved to snorkel. It was her passion. But there was one occasion when she was in the ocean and she lost track of time. When she finally surfaced, she noticed that she had drifted a long way from where she started. She was frightened and began to quickly swim back to where the rest of her friends were. She made it back to shore exhausted and noticed that she started to feel unstable. When she arrived home, she was still unsteady. After a few days, she decided to go to an Ear, Nose and Throat specialist thinking she may have something in her ears. She was given a few prescriptions and went on to visit several other specialists who told her that her dizziness was due to an inner ear imbalance. Within six months, she was no longer able to work. I saw her three years later when she had all but lost hope of ever getting better. But within minutes of treating her, everything changed. She began to stretch out her arms, and shrug her shoulders repeatedly, as if she was testing her upper body to see if anything would hurt. She turned to me and smiled. "My dizziness is gone," she said with a surprised look on her face. While examining her ear with an electrical detection device, I found electrical activity in the area that corresponds to the stellate ganglion, which is a collection of nerves on either side of the spine near the base of the neck/shoulder area. This area was also very sensitive when touched. By treating this area on the ear with gentle pulses of light and a tiny ASP (French) needle, her symptoms disappeared. She was fine. No more dizziness. She was able to return to work and is totally normal to this day. Her diagnosis was simple: A compression of the stellate ganglion.

The second story is one that is far more unusual, but it demonstrates the broad range of use for auricular medicine. A twenty-six-year-old woman was referred to me due to chronic pain in both her jaws and ears. She had seen several specialists and was being treated for recurrent ear infections. I ruled out Temporomandibular Joint (TMJ) disorders as well as muscular pain. I then examined her ear with a gentle pulsed light and noticed that the opening to her ear, where she complained of recurrent infections, had an unusual response to certain frequencies of light.[1] With no preconceived notions, I scanned her entire ear and found an area that responded to the pulsed light in the same way that her ear opening did. It was a match. This area on the rim of the ear happens to correspond to the reproductive organs. I asked her when she first noticed the jaw pain and ear infections. She said about two years ago. I then asked if she had experienced any events that may have led to the recurrent infections and pain. She said no, and then she remembered something. About two years ago, she

[1] Her pulse became momentarily stronger.

had a birth control device placed, and it was then that she first noticed the pain in her jaw and the onset of ear infections. I recommended that she make an appointment with her doctor right away. The next morning, she had the device removed, and she later called to say that the pain she had been experiencing for the past two years was gone. She returned to normal. Her ear infections completely healed and haven't returned. The diagnosis, like in the first story, was simple, and in both stories, it was the ear that guided us to the needs of the body.

A well of living water from which healing will spring up speedily

## A healing language

Auricular medicine is an approach to health that has age-old ties dating back to 3000 BC, and it's far from the fixed ideas that most of us have about our body and how we can best maintain it. Historical writings speak of a *well of living waters* and the *light of life* from which *healing shall spring up speedily.* Beautifully spiritual, these words also emphasize the life-giving power of water and how light is endowed with power essential to life. Yet, time and again, we ignore the wisdom of our ancestors and choose to follow the latest trends and fads while being fed statistics that often hide reality at the individual level. Nonetheless, each of us is unique with a particular set of genes and history. And for those of us who have lost our health and want to restore it, it's important to implement a more personalized approach; one that uses a remedy that our body is asking for. In addition to the treatments that we are familiar with, such as drugs, surgery, or exercise, *therapy should also be considered a language, a healing language.*

To understand this language, it's important for us to realize that every form of life we know of depends upon frequencies. Frequency is the number of times an event repeats itself in a given period of time. It's often measured by the number of times an object vibrates, or moves back and forth, in one second. The light we see and the sounds we hear are in the visible and audible vibratory range. Yet, there is a symphony of frequencies that surround us that we simply can't sense.

## Frequencies affect our health

It's been well illustrated that our health can be affected by frequencies that come from either beneficial or harmful thoughts, emotions or words, or from more visible events such as a physical injury or over exposure to sunlight. Although some of us are more sensitive than others, we can all relate to the feeling of being stressed out or even being hurt by something that was either said to us, done to us, or by the mere anticipation that something bad is going to happen. *Regardless of its origin, when we are exposed to something unhealthy, the seeds for disease are sown.* Many times, the invisible frequencies that are not within the healthy range of our body may go unnoticed for years and only become visible when mental, emotional, and physical symptoms erupt. Thankfully, due to the discoveries of a gifted neurologist, Dr. Paul Nogier and his son, Dr. Raphael Nogier, *state-of-the-art therapies have been developed to feed our cells their natural healthy frequencies that will give them the opportunity to once again vibrate with health.*

A discoveries' worth depends on the people to whom it will help and Dr. Nogier's discoveries are truly a treasure that can be shared by everyone.

## The largest and most sophisticated pharmacy in the world

Today, we are being inundated with direct-to-consumer pharmaceutical drug advertising. As a consequence, this influences us into thinking that using a particular product is a normal, ordinary, natural part of life; an everyday thing to do that everybody is doing, even when the list of side effects can be worse than the disease itself. *What we aren't being told is that our body is actually the largest and most sophisticated pharmacy in the world.* Scientists estimate that each of our cells undergoes about one hundred thousand chemical reactions per second, and this is without the aid of prescription drugs.

Our body draws from the environment and uses all of its surroundings to build the substances it needs to keep it alive and restore its health in moments of weakness. However, from time to time, when we become injured or ill, we do need assistance. *Helping the injured area return to its natural, healthy frequency can accelerate this healing process.* A practitioner who understands and can properly apply frequency diagnosis and treatment to a particular ailment can accomplish this. Areas on the ear that correspond to an injured area on the body can be stimulated by either light or electrical frequency, or tiny acupuncture needles to help it regulate and repair itself and return to a healthy condition.

## Our skin and our health

Our nervous system has antennae in every area of our body to both receive information from the outside and within and to help produce substances that transmit information to ensure that everything is working well. The cells that make up both our nervous system and our skin are derived from the same basic "stem" cells that form early in our development. This *intimate connection between our nervous system and our skin plays a very important role in our health.*

Our skin is much more than just a physical barrier to the outside world. In addition to helping regulate body temperature and being a main source of vitamin D, as well as detecting different sensations such as pleasure or pain, it does something remarkable; something that we have only recently discovered. *Our skin is capable of receiving and interpreting light*, a quality known as photoperception. This means that not only is it sensitive to light but it is also able to *transmit information coded in the light to our spinal cord and brain with the purpose of restoring our health.*

Dr. Paul Nogier identified seven frequencies[2] that a healthy person's body will naturally respond to. *The response to the frequencies of pulsed light is measured by a subtle increase in pulse strength.* His research shows that the *application of these frequencies of light helps*

[2] A frequency can be defined as the number of times per second a light flickers on and off.

*to bring our body back to its natural healthy state.* We also know that every ailment or painful area in our body has what I like to refer to as its own personal *light signature,*[3] which reveals information on how to best treat it. Just as each of us is unique, the treatment we receive should also be unique and individualized. Thus, *our skin plays a key role in the regulation of the nervous and immune systems, which in turn communicate with and help regulate every other system in the body.* This personalized approach to healing can help us with a wide variety of health issues and concerns; often after all other medical interventions have failed. Our body is telling us what it needs, but we need to understand what it's saying; we need to recognize that *there is a unique healing language that has been imprinted in each of us.*

3 A unique response to different frequencies of light.

## Chapter 2

# Words Matter

Words are powerful. When you speak, you give life to what you are saying. The words you say can put a smile on someone's face or bring tears to another one's eyes. But is there truth to the belief that words, thoughts, or messages that are either spoken, communicated nonverbally (i.e., body language), or written can never truly harm you? And that only a physical attack can injure or scar you? The answer is no. The truth is that *the words you say and believe can be likened to seeds and these seeds will give life to your words. By saying positive words, you will plant the seeds of health and healing.* Whether you are aware of it or not, the words you say are planting the seeds for your future. They can have a wonderful, positive, healing action when you say something positive such as, “I will get better and be able to do those things I dream of doing”. However, if you chose to talk about yourself in a negative way, how can you possibly expect to live a positive, healthy life? If you do not like how you are feeling due to a physical and/or emotional ailment that has lead to chronic pain, it will be important to begin planting some different seeds. Instead of saying, “I am never going to get better and no one will ever be able to relieve my pain,” begin saying, “My health will get better and will continue to improve with each passing day”.

Your words reveal what’s in your heart. And in the same way that a tree is known for the fruit it bears, your thoughts, feelings, and desires are known by the words you speak. When you keep planting these positive seeds, you will eventually enjoy the fruits of health. Simply said, you can change your health by changing your thoughts.

Words can be likened to seeds

## Research that shows how words and thoughts affect your brain

There's been groundbreaking research by several scientists that has focused on changes that occur within your brain when you are exposed to either negative or positive words. These changes are measured by using functional Magnetic Resonance Imaging (fMRI) technology, which is a machine that looks like a large donut-shaped magnet and it can take a video of our brain activity.

Hearing negative words affects your brain activity

The research is clear that exposure to negative words will cause a sudden release of dozens of stress-related hormones, which quickly interfere with the normal functioning of your brain. The longer you think about negative words or events, the more you damage important area's that regulate your memory, feelings, and emotions. This leads to disturbances in your sleep, appetite, overall happiness, and your ability to manage pain.[4] This is why worrying is so harmful; it causes the release of stress hormones such as cortisol. When it becomes chronic, worrying changes brain function even down to the level of your DNA.[5] The harmful effects of worrying have been known for thousands of years as historical writings have highlighted the importance of not worrying about tomorrow, for tomorrow will worry about itself and that each day has enough trouble of its own.

---

4 A. Talasovicov et.al., Regul. 2007 Nov; 41 (4): 155–62.

5 Pedersen, T. 2012. Stress Increases Risk of Mental, Physical Illness by Altering Genes, Psych Central.

While your brain responds quickly to negative words or thoughts, it hardly responds at all to positive ones.[6] This is because good words and feelings are not a threat to your survival.[7] Positive words and thoughts, which stimulate the motivational centers in your brain, help you build up resilience when you're faced with problems.[8]

Based on research, it's best for your health and for the management of pain to choose your words and thoughts carefully. This interferes with your brains tendency to be negative and as recent research has shown, the mere repetition of positive words and thoughts will actually turn on specific genes that will lower both your physical and emotional stress.[9] When you dream up a minimum of five positive thoughts to each negative one, you tend to function at your best.[10]

---

6 Kisley MA, Wood S, Burrows CL. Psychol Sci. 2007 Sep; 18(9): 838–43.

7 Smith NK, Cacioppo JT, Larsen JT, Chartrand TL. Neuropsychologia. 2003; 41(2): 171–183.

8 Cohn MA, Fredrickson BL, Brown SL, Mikels JA, Conway AM. Emotion. 2009 June; 9(3):361–368.

9 Dusek JA, Out HH, Wohlhueter AL, Bhasin M, Zebini LF, Josesph MG, Benson H, Libermann TA. PLoS One. 2008 July 2;3(7) e2576.

10 Fredreickson BL, Losada MF. Am Psychol. 2005 October; 60(7):678–686.

## Just breathe

In addition to the effects that your words and thoughts have on brain activity and health, the effects of relaxation and proper breathing are also significant to your ability to manage pain and overall health.

Simply by relaxing your posture and taking the time each day to breathe properly, you can change pain signals by reducing the demands on your body and mind and by increasing oxygen intake and blood flow. *Breathing is like a bridge that connects your body to your thoughts. By breathing properly, you can help to reduce both physical and emotional pain.*

As we all know, breathing occurs naturally in three stages: Inhale, exhale and pause. But what many of us don't know is that there is a right and wrong way to do this with regards to relaxation, reducing overall stress and pain relief. Breathing through your abdominal area, known as diaphragmatic breathing, is recommended. Your diaphragm is the most efficient muscle for breathing and is much more vascular (more blood vessels) than the muscles you often use in your chest. With so many blood vessels in the diaphragm, your body can absorb more oxygen with each breath.

I recently met with Dr. Charles C. Carlson, who has been named the 2015–16 Distinguished Professor in Arts and Sciences at the University of Kentucky. He has published over 115 research papers on helping people to better manage their responses to pain. His research has concluded that *diaphragmatic breathing* improves your circulation, provides more oxygen throughout your body, and reduces the overall demands on your body to provide *pain relief. The ability to control breathing is the closest thing that you have to a switch that can turn on and off (regulate) your nervous system.* It may take some practice, but in time, using the diaphragm to breath will become natural, just as it is for children.

## Diaphragmatic breathing

To practice diaphragmatic breathing, also known as belly breathing, place one hand on your abdominal area and practice breathing to see your hand rise (as you inhale) and fall (as you exhale). The pattern that I recommend to my patients is as follows: Inhale and count "two, three" then exhale (do not control, just let it go) and then pause "two, three." In other words, the sequence is to rise, release and rest. To slow your breathing rate, it is best to lengthen the pause or rest phase. The goal is to reduce your breathing rate to three to seven breaths per minute, which will decrease stress, boost your immune system, and reduce your pain. Research by Dr. Carlson has demonstrated that it's important to take brief breaks each day to include a relaxed and comfortable postural position, and diaphragmatic breathing. Begin these relaxation breaks for a period of five minutes one to two times daily and gradually increase to twenty minutes one to two times daily, which according to research and clinical observation, will lead to improved management and control of both physical and emotional pain.

Dr. Kalfus catching a breath in the Republic of Vanuatu in 2011

# Chapter 3

# Ears Do More than Capture Sound Waves

Sound is so important to us. It's likely to be the first thing we experience when we wake up in the morning and the last thing we hear at night when we listen to our heartbeat and drift gradually into a soundless world of sleep. Even though the most recognized purpose of our outer ear is to be a wave catcher and to deliver the sound it captures to the middle and inner ear so that our brain can interpret the waves, our ears are designed to do much more.

## Different uses of ears

While some species have relatively large ears for echolocation (i.e., bats) to provide them with hearing abilities far greater than most land animals, other species have different uses for their ears. The African elephant's ears, for example, are over twice as large as the Asian elephant's ears and have a different shape, often described as being similar to a map of Africa. Elephants use their ears to communicate visually. Flapping their ears can signify either aggression or joy. When alarmed or angry, they spread their ears, bringing them forward and fully extending them. Their ears also control body temperature. By flapping their ears on hot days, the blood circulates in the ear's numerous veins and return to the head and body about 9 degrees fahrenheit cooler. Indian elephants, on the other hand, have much smaller ears because they live in moist jungles in Asia. Desert animals, such as the desert fox and the desert hare, also have larger ears as compared to their non-desert relatives, the arctic fox and the arctic hare. The purpose of these differences in ear size is the animals need to keep their body temperature within certain boundaries, even when the surrounding temperature is different. The thin skin of the highly vascular ears provides an important source of heat loss for desert animals while animals in freezing cold temperatures need to conserve internal body heat.

For human beings, our ears have a significantly high number of points that are both electrically active and electrically inactive. Each of these points is made up of a collection of blood vessels, nerves, and immune cells that are here for a reason; they communicate with and help regulate all of our body systems to include our body temperature, repair and maintenance, and pain management. It's evident that our outer ear is much more than just a wave catcher.

# Chapter 4

## A Key to Health and Aging

The earth is a watery place. About 71 percent of our planet's surface is covered in water and about 96.5 percent of all earth's water is held in the oceans. The remaining 3.5 percent of water exists in the air as water vapor, in rivers and in lakes, in icecaps and in glaciers, in the ground as soil moisture and in aquifers, and even in you. On average, we are made up of about 75 percent water (at a molecular level, our bodies are more than 99 percent water) and keeping our body hydrated is essential to our health. *Water found in healthy human cells*, however, *is not*

*ordinary.* Rather, it is *highly organized.* And *the key ingredient to water molecules becoming organized is sunlight* (electromagnetic energy), whether it is visible light, infrared (the most powerful) or ultraviolet.

Among the total spectrum of solar rays coming from the sun, the greatest amount of the sun's energy output is in the infrared segment of the electromagnetic spectrum, and it is the *far infrared waves that are the safest and the most beneficial.* This band of light is not visible, but it can be felt as heat that penetrates deeply into our body with a uniform warming effect. According to Dr. Mu Shik Jhon, the world's leading authority on water science, it is *how water molecules bond together that may be a key to health and aging.* In particular, Dr. Jhon has found that *living organisms prefer a six-sided ring-like organization of water molecules rather than random clusters of water molecules* found in most tap and bottled water that are often too large to freely move into cells. He summarized his forty years of research into a simple theory that defines aging as a loss of hexagonal water from organs, tissues, and cells, and an overall decrease in total body water. As we age, our water percentage by weight decreases from approximately 80 percent at birth to less than 50 percent with advanced age. This loss of water also correlates to the loss of hexagonal water, which is close to 100 percent at birth and steadily drops as we age.

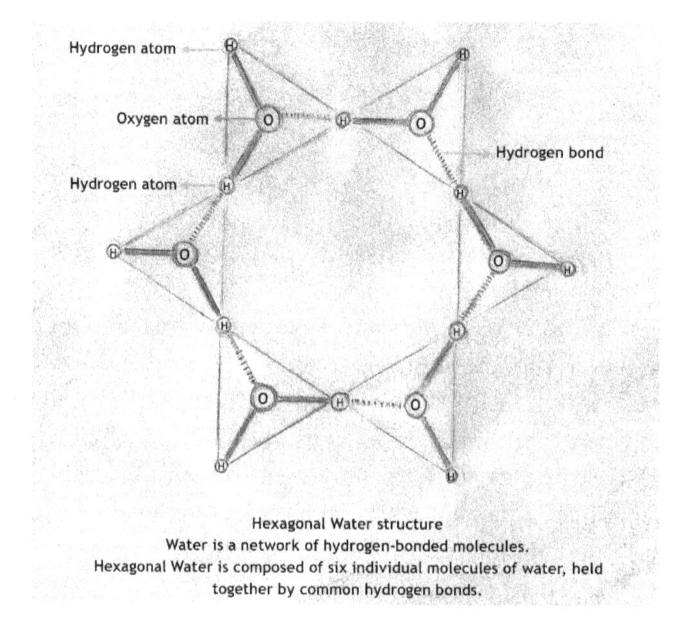


Hexagonal Water structure
Water is a network of hydrogen-bonded molecules.
Hexagonal Water is composed of six individual molecules of water, held together by common hydrogen bonds.

When clusters of water molecules are too large, they must be restructured within the body before they can penetrate the cells, and this is a time-consuming and energy-consuming process. When clusters of water are in a six-side ring-like organization, or hexagonal shape, which is found *naturally* in snow water, water in fruits and vegetables, and cold water from deep wells and pristine streams, there is rapid penetration into the cells. The scientific consensus of researchers familiar with cellular water structure and Nuclear Magnetic Resonance (NMR) is that the water environment surrounding unhealthy cells is less structured and thus able to move more freely than the water environment around healthy cells.

According to Dr. Jhon, who has published nearly three hundred scientific papers on the subject, *hexagonal water forms a protective layer immediately around healthy proteins.* This same type of protection does not exist around unhealthy proteins in the body where pentagonal (five-sided) water clusters are found. Water surrounding normal DNA[11] was found to be highly hexagonally structured, which acts to stabilize the helical structure of the DNA, and forms a layer of protection against outside influences that could cause malfunctions or distortions within the DNA. Also, within the core of each DNA double helix is a column of organized water clusters, which allows it to freely pass through cell walls delivering oxygen, nutrients, protein chains and enzymes; it also removes accumulated toxic materials in a way that unstructured, loosely bonded water (i.e., tap, rain, or mineral) cannot.

Science has demonstrated that approximately 90 percent of DNA function involves electrical, magnetic and light reception and transmission. *This optimal function, underlying every aspect of health, delaying aging, and slowing down virtually every disease process, almost entirely* depends on the arrangement of the water molecules in the core of our DNA strands. Within each living cell, each molecule of protein is surrounded by thousands of molecules of water, which means that DNA cannot be looked at as a freestanding molecule; rather, it is an integral part of a huge organized water

---

[11] DNA is a molecule that carries most of the genetic instructions used in the growth, development, functioning and reproduction of all known living organisms and many viruses

cluster. The communications between these clusters is what we call life. No life processes can take place without water. The structuring of cellular water is critical to the healthy functioning of every cell. Cells surrounded by less-structured water may be weaker and more prone to disease and genetic change.

The therapy discussed in this book uses the same light and electrical frequencies that our DNA uses to function, and it serves to stimulate, restore, and maintain health; this gentle healing language, or treatment, is applied at specific points on your ears and transmitted through a network that is comprised mostly of water. What is most important here is to know that the water found in healthy human cells is not ordinary. Rather, it's highly organized and the key ingredient to water becoming organized is light, which is central to auricular medicine.

# Chapter 5

## Chronic Disease May Be a Food-Borne Illness

Over the past century, the world has changed. Our diets have changed to. In my own lifetime, we've moved from gathering together for a traditional family "supper" as I routinely did with my parents, to a more rapid and often solitary processed meal that today is far too rich in animal proteins and fats, and far too poor in minerals. Just as our eating habits have changed, so too has the landscape of our diseases.

For all of the good that the discoveries of antibiotics, immunizations, and antiviral drugs have done to decrease the spread of once dire infectious diseases, there has also been a *steady rise in degenerative or inflammatory diseases, which is likely linked to what we are eating.* Many scientists mistakenly predicted that resistant strains to the "magic bullets" (or antibiotics) that had been discovered wouldn't occur or much less spread. Yet, bacteria and viruses have been around a long time, and a key to their success is their ability to adapt to changes in the environment. When they reproduce within a matter of minutes or hours, those changes come quickly. Thankfully, our immune system is often up for such challenges.

Along with the changes in medicine and in the organisms that threaten our health, the world in which we live has also changed. In addition to tremendous advancements in technology, there have been sociological changes, changes in our diet, as well as changes in the types of diseases that now fill our landscape.

The sociological changes include the breakdown of the classical family, the increase in individuals living alone, the exodus of people from the countryside to urban living,[12] the increase in women in the workforce,[13] the occurrence of foreign populations,[14] the rise of supermarkets, and advertising in media.[15]

The technological changes are many and include alterations of plant genetics (mainly in bread and wheat cereals), the use of insecticides (upsets ecosystems and pollination by bees), agricultural mechanization,[16] changes in irrigation,[17] and overworking the land (we no longer let the land rest every three years to replenish the minerals in the impoverished farmland).

---

12 Adopting to alternative foods.

13 Reduced breastfeeding.

14 Introducing foods and preparations not previously known by local populations.

15 Targeting children and encouraging sweet and fatty foods.

16 Resulting in monocultures of a single crop over huge area's that are chemically very different.

17 Producing chemically inconsistent crops.

In addition to the social and technical changes in agriculture, we have also seen changes in food preservation,[18] food presentation,[19] and the use of chemicals in livestock.[20]

As a result of all of these changes, the food that we consume today is a lot different from what our parents enjoyed. Today, our diets are much higher in protein and lipids[21] and much lower in minerals as compared to just a half-century ago. The decline in complex carbohydrates (three or more sugar molecules strung together) is in favor of beet sugar and cane sugar (one or two sugar molecules that are quickly digested) and the consumption of meat, fish, and dairy has also increased. In the meanwhile, breastfeeding had dropped considerably after the 1940s–1970s but has since had an upward trend. Additionally, the rates of physical activity have steadily declined.

All of these changes are reflective in the types of diseases that we see today. With the discoveries of antibiotics and immunizations, diseases of bacterial origin, which are the primary sources of epidemics, have been controlled. Yet, at the same time, we have seen the rise in viral diseases (such as hepatitis and AIDS). But *the most significant change that characterizes today's landscape is the explosive increase of chronic (noncommunicable) diseases.* In France, for example, chronic diseases have increased by 73.5 percent just between 1994 and 2004. Chronic diseases constitute the main cause of death in the EU, accounting for 77 percent of the total. *The most important are heart disease, stroke, and cancer.*[22] In the United States, *obesity* in adults and children has more than doubled since the 1960s, and this increase is seen an all age groups, both sexes, all racial/ethnic groups, and in all states.[23] *Chronic Respiratory Diseases,* such as bronchitis and emphysema have increased by 50 percent during the past thirty years and continue to rise.[24] There has been a sharp rise in

---

18 Dehydration, refrigeration, smoking, and salting.

19 Use of dyes, waxes, flavor enhancers, and advertising to encourage consumption.

20 Antibiotics and steroids.

21 Two times higher than fifty years ago.

22 Health in the European Union, Observatory Studies Series No. 19, 2009.

23 National Center for Health Statistics, United States 2010.

24 Anderson HA, Chronic Respiratory Diseases, American Public Health Association 2010.

*autoimmune disorders* over the past fifty years to include rheumatoid arthritis, type 1 diabetes, multiple sclerosis, lupus, Hashimoto's thyroid disease, celiac disease, and asthma. There are more than ninety autoimmune diseases. Since the 1950s, celiac Disease alone is up fourfold, lupus rates have tripled, and type 1 diabetes has soared—up twenty-three percent in the past decade alone.

Decades of research have provided us with *three basic ingredients that every autoimmune disease has: A genetic predisposition, an environmental trigger, and intestinal hyper- permeability*, which is also referred to as a "leaky gut".[25] In particular, a protein was discovered that regulates the permeability of the digestive tract. This protein, called Zonolin, opens spaces between cells, allowing some substances to pass through while keeping harmful substances out. Some people produce excess amounts of this protein, which pries apart the cells that line the intestines and allows toxins, bacteria, viruses, fungi, and undigested bits of food into the bloodstream—hence the term "leaky gut." Confronted with a steady stream of invaders, the immune system makes what are called T-helper cells, which speed up the response and also *trigger autoimmunity. The surge in autoimmunity paralleled the growth of consumer products made by plastics, artificial fillers, and synthetic dyes.*

There are more than eighty thousand chemicals used in commercial products, and less than fifteen thousand of these have been safety tested. According to consumer advocates at the Environmental Working Group, the average person is exposed to more than 126 chemicals before leaving the bathroom in the morning (shampoo, toothpaste, soap, etc). Women make up 78 percent of autoimmune sufferers, and their personal care products may be a leading cause because of the endocrine disrupters (phthalate esters and parabens) found in lotions, perfumes, and sunscreens. What's important here is to realize that each of us is unique with a particular set of genes and history and that each of us has his or her own reaction to food and various chemicals. We shouldn't rule out that the explosion of chronic disease is linked to the explosion of food sensitivities. *There's likely a link between autoimmune disease and food sensitivities.* However,

[25] Discovered in 2000 by Alessio Fansano, M.D.

the subject of food allergies is extremely complex. It's been said that understanding food allergies is as clear as a bottle of black ink.

There is a likely link between food sensitivities and autoimmune disease

## How do we become allergic to food?

*In general, under healthy conditions, no one should be allergic to food.* Our immune system relies on specific cells called lymphocytes (a type of white blood cell) that will defend us against bacterial, viral, and parasitic attacks. These lymphocytes detect foreign proteins, and only proteins. They *cannot* recognize sugars or fatty acids (i.e., omega 3). Thus, they detect viruses and bacteria that have proteins within them. In fact, these immune cells don't recognize the virus and bacteria as such but the proteins that compose them. The food we eat consists of water, minerals, carbohydrates, lipids, and protein. *The dietary proteins are foreign to our body. They are tolerated in the digestive tract but considered an intruder in the bloodstream.* Normally, food proteins do not enter into the bloodstream. Proteins that are absorbed are degraded into amino acids in the digestive tract. When they arrive in the small intestine, most of them are denatured (i.e., their structure is changed). There is also a barrier in the digestive tract that blocks the passage of proteins and no food proteins can be found in the blood. Thus, when we eat, *protein normally does not cross the intestinal barrier. If food protein does cross this barrier, it will enter into the bloodstream and be identified as an intruder by the immune system. This is the beginning of the allergic process.*

In summary, food allergies can only occur if a food protein passes the intestinal barrier and is found in the blood.[26]

## The first four to six months of life are unique

There is a time in our life when there is no functional intestinal barrier and this is during the first four or six months of life. Until then, the intestinal barrier is somewhat open. This allows infants to absorb mild proteins, some of which will protect against infections and are known as immunoglobulins. Thus, a mother can transmit to her child all the antibodies she has made during her lifetime. Importantly, this permeability can become a weakness if the child is not breastfed by his mother. If a baby is fed milk powder (cow milk), the proteins of this milk will likely cross the intestinal barrier and create sensitivity to cow's milk. If the mother cannot breastfeed, she should be advised to give her infant only hypoallergenic milk. The more diversified the infant feeding is during the first six months of life, the greater the risk to see food allergies develop. Therefore, avoid feeding an infant with products other than hypoallergenic milk if they cannot be breastfed because *when infants ingest a foreign protein in their first four to six months of life, they may eventually develop an allergy to the protein.* Importantly, nursing mothers must be advised to avoid allergenic products such as ground nuts, exotic fruits, and limit the intake of dairy products and eggs because the baby can develop an allergy.

Additional factors that affect the permeability[27] of our intestinal lining include anti-inflammatory drugs, antibiotics, birth control

---

[26] Note that adding glutamine to your diet is the best known way to decrease hyperpermeability in the digestive tract.

[27] How easily things pass through a barrier.

medication, sustained exercise,[28] alcohol overuse, during times of infections (flu or bacterial infections), and following vaccinations.[29]

When a dietary protein crosses the digestive tract barrier and enters our bloodstream, it becomes unwelcome and our immune system will respond in seconds, minutes, hours, and sometimes days. There are three types of possible reactions to a food allergy: Type I reactions are the most frequent and appear quickly after ingestion of food. Immune cells will bind to the food protein causing a release of potent chemicals such as histamine. This chemical will alter the permeability of the blood vessels and cause several noticeable signs such as hives and a drop in blood pressure. A blood test usually reveals a high level of food-specific immune cells known as Immunoglobulin E (IgE). Type I reactions are in genetically predisposed individuals and the treatment usually consists of antihistamines and bronchodilators. Type III reactions occur a few hours after ingestion of food. The undesirable proteins are "captured" by immune cells called Immunoglobulin G (IgG) to form large structures that circulate in the blood and tend to settle in specific organs or obstruct small arteries. Treatment for Type III reactions usually includes anti-inflammatory medications. Type IV reactions occur with a delay of twenty-four to seventy-two hours and depend on an immune cell called a T-Lymphocyte. Besides the long time required for the onset of the clinical signs of an allergic response, there is almost no role of individual predisposition.[30] The treatment for a Type IV allergic response is usually a corticosteroid along with other immunosuppressive medications. There is a Type II allergic reaction, but it is NOT related to food. Type II reactions occur within minutes or hours and is mediated by Immunoglobulin M (IgM) and Immunoglobulin G (IgG). Treatment usually involves anti-inflammatory and Immunosuppressive medications.

---

28 Muscles use oxygen reserves at the expense of intestines and cortisol levels rise.

29 Some vaccines contain food or animal proteins in which the virus or bacteria were grown.

30 Person doesn't have to have any previous exposure to the protein that caused the reaction.

*The food allergies described in classical medical books are mostly* IgE mediated (Type 1) and *immediate.* The other types of food allergies are mentioned but seldom given any attention by the authors. Their diagnosis is difficult because the laboratory tests are not reliable. *Signs of hypersensitivity will disappear when the dietary "protein" is excluded from the diet and will reappear with a minimal food intake suggesting that they are linked to the immune processes.*

Food allergies have been implicated in many diseases for many years and the signs are plentiful. They include general signs (fatigue, weight disorders, and sleep disorders), neurological signs (depression, attention deficit, headaches, and anxiety), skin disorders (eczema and sensitivity to the sun), cardiovascular signs (hypotension and increased heart rate), intestinal signs (constipation, diarrhea, and pain), rheumatologic signs (unexplained joint pain and joint inflammation), gynecological signs (breast pain and premenstrual syndrome), ear, nose and throat signs (allergic runny nose), endocrine signs (inflamed thyroid), and stomatological signs (canker sores).

*The main foods that contribute to delayed food allergies are cow's milk, beef, wheat, corn, potatoes, tomatoes, chicken eggs, soybeans, and citrus fruits.*

Regarding allergy to dairy products, I often wonder why Western civilization is so enthusiastic for dairy products and the *misconception that milk is essential to health.* While it's almost impossible to escape the advertising telling us that dairy is good for everything, we need to realize that only one-third of humanity has access to milk and its derivatives. The vast majority of the over four billion people without dairy products are neither malnourished nor sick. Also, the advertising is not always truthful. For example, *contrary to popular belief, dairy products have no positive effects on osteoporosis. No scientific study has shown that a milk diet prevents osteoporosis.* It's not how much calcium we eat that's important. Rather, it how much calcium we can prevent from leaving our bones that counts. In countries having the lowest amount of milk consumers, the rates of osteoporosis are lowest. On the other hand, countries having the largest amount of milk consumers have the highest rates. While it's true that in osteoporosis, there is less calcium in the bones, it's a little elementary to state that since milk contains calcium, it's good for osteoporosis. This disease is very complex, and its mechanisms are not fully understood.

For several years, my colleagues and I have suggested to patients (primarily women) to have *diets without milk proteins.* This allowed us to observe the disappearance of fatigue, chronic inflammatory diseases, intestinal disorders, neuropsychological manifestations (migraine headaches and mood swings), skin disorders (eczema), rheumatologic joint pain as well as gynecological manifestations (breast pain and ovulation disorders). These "*avoidance diets*" are the only treatment of allergies to milk proteins and their derivatives that I recommend. Sometimes, it's necessary to exclude beef and veal due to cross allergies. Here's a typical example of a patient I treated that is allergic to cow's milk. Referred to me due to frequent headaches and sore joints including her TMJ, a 24 year old female patient also told me that she has chronic fatigue and is generally anxious. She stated that since childhood, she has always felt tired and was often at the doctor. I asked if she was bottle-fed as a baby, and she said yes. She was fed with milk powder. When she was very little, she had intestinal colic, along with frequent ear, nose, and throat infections. She was often prescribed antibiotics for her ear infections as well as cortisone. As a teenager, she was advised to take calcium supplements and was told that she had low blood pressure. As a result, she often hyperventilated when she exerted herself. On her first visit with me, she arrived with a migraine headache and was very fatigued. Her blood pressure was also low. I asked her if she drinks milk. She answered by saying that she didn't like milk at all but eats a lot of dairy to include yogurt and cheese. After treating her for her migraine with auricular therapy, I asked her to remove milk and all milk-related foods from her diet. I recommended she use plant-based substitutes like almond or soy and to avoid foods with milk protein. A month later, she returned for a follow up visit and she stated that her pains were all gone. She was no longer fatigued, her headaches haven't returned, her joints no longer ached and she felt better overall. This was likely an allergy to milk proteins. *Whenever I see a patient with a chronic condition, I look to rule out food allergy.* Importantly, I stress to my patients that it may take several weeks or months for the "porosity" of the intestines to heal and during this time, food proteins other than the one's causing the food allergy can cross the intestinal barrier and cause a secondary allergy. We must be patient and wait for the intestinal mucosa (lining) to heal before adjusting or discontinuing the avoidance diet.

As mentioned above, our immune system relies on cells called lymphocytes. In adults, there are approximately 1012 types of lymphocytes, which is truly amazing. Each one can recognize one specific protein. This is like having a police officer that's only able to seize and capture one specific thief. But, there are trillions of different proteins possible. So in order to have trillions of different possible receptors for these proteins, the lymphocytes through recombination of their genes, produces billions of cells, each with a different recognition system. Therefore, to avert the possibility of protein attack, our body has in reserve a specific anti-protein lymphocyte for each possible protein. This is a miracle.

In summary, as our diets have changed due to sociological changes, technology changes, agricultural mechanization, insecticides, food preservation and presentation, overworking the land, and the use of chemicals in livestock, the types of disease that we see have also changed. The most profound change has been in the sudden rise of chronic diseases such as chronic respiratory diseases and autoimmune diseases. *With chronic disease, I always want to rule out a food allergy by simply employing an avoidance diet.* The hyper-permeability or "porosity" of the intestinal lining will need to heal to prevent food proteins from entering into the bloodstream. When proteins do enter the bloodstream, they are seen as intruders by our immune system and this is the beginning of the allergic process.

Having been in practice for over twenty-five years, I've learned that helping people heal is truly an art. It's an art of compromise, of balance, of common sense, and of measure.

## Chapter 6

# Everything Seems to Be Turned Upside Down

Imagine for a moment that everything in the world seems to be reversed from what we are used too. Here, the sky would be below us and our feet would be firmly planted to the ground above. A water droplet from an icicle would gently fall up and the warm glow of a candle would point down. Even a chandelier would hang straight up and point away from the ceiling below. Although this place sounds strange to us, it wouldn't be so unusual to our brain. In fact, this is the way that our brain sees the images from our eyes before it flip-flops them to be right side up again. And in the same way that our brain see's the world, our ear also has an upside-down mapping of our body[31] with our feet represented at the top of the ear and our head represented at the bottom of the ear. However, unlike our brain, which inverts the images it receives to be right side up again, the upside-down body image on our ear remains inverted.

Sometimes, upside down seems to be right side up

## An upside down map of our body is found on the ear

There is an intimate connection between our ears and our brain that isn't found anywhere else in our entire body. In addition to having direct and indirect connections to our abdominal organs, thoracic (chest) organs, musculoskeletal system (muscles and bones), the nervous system (brain and spine and all the nervous tissue that connects to the spine), and our skin, there's another *unique connection between our ear and our brain*; it's called inversion or upside-down orientation. Most of the time, our entire body, from head to toe, is represented on our ear in an upside-down position.

---

[31] That is an image of our entire body that is projected onto our ear.

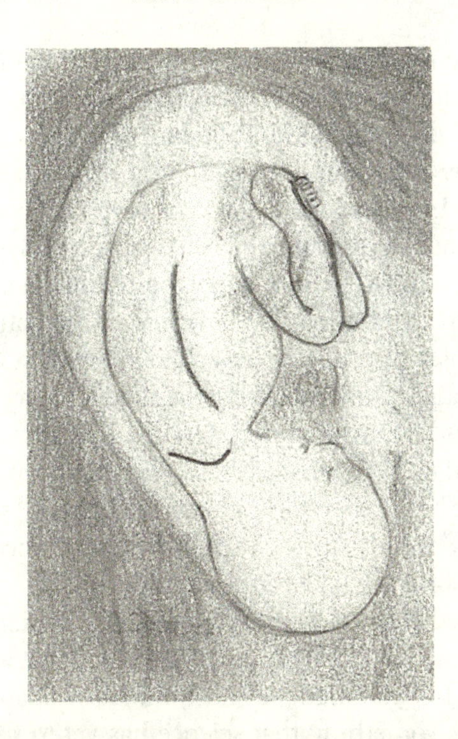

Such an *organized arrangement is unique to the ear and is found nowhere else on our entire body.* This means that the top of our ear corresponds the bottom of our body (i.e., our feet) while the bottom of our ear corresponds to the top of our body (i.e., our head). This upside-down map of our body is referred to as *Somatotopic Inversion* and makes more sense when we realize that our brain sees the world upside-down.

## We see the world upside down

A famous experiment was conducted in the late 1800s, in which George Stratton, a psychologist, published "Vision without Inversion of the Retinal Image."[32] The reason why it is called "without inversion" is that due to the quirks of optics, the image that hits the retina of the human eye is upside down. The brain automatically inverts it, so we see the world the way we do, which is right side up. Glasses used

32 Stratton, George M. Psychological Review, Vol 3(6), Nov 1896, 611–617.

in this experiment made it so that Stratton, for the first time in his life, saw a right side up image of the world. Not surprisingly, his brain turned it upside down again, and he blundered around and crashed into things. After a few days, though, Stratton was able to adapt and work as normal. In one instance, he wore reversing glasses for twenty-one and a half hours a day over the course of three days and found no change in his vision. After removing the glasses, normal vision was instantly restored without any disturbance in the natural appearance or position of object. On a later experiment, Stratton wore the glasses for eight days straight, without taking them off. On day four, the images appeared upside-down or inverted. On day five, however, the images appeared upright until he concentrated and they then became upside-down again. The brain, therefore, adjusts to changes[33] so that the world appears "normal." Additional research by Professor Theodore Erismann and Ivo Kohler was conducted in the middle of the twentieth century also demonstrated that after several weeks of wearing goggles, which "reversed a person's world," we adapted to visions that were topsy-turvy or backwards. This automatic, almost effortless adaptation to visual inconsistencies is something that science has yet to fully understand. We know that we see things upright even though the image formed on the retina in the back of our eye is an upside-down, inverted one.[34]

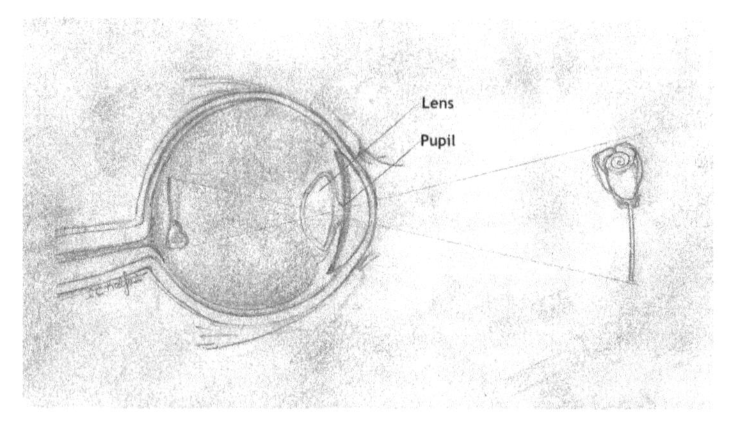


The Image of the rose is inverted on the back of our eye

[33] In this case the inverted images.

[34] Retina is Latin for "net" and it is a light sensitive layer in the back of the eye that acts like the film in a camera.

As with all of our senses, we can, for example, adapt to sounds and tend to distinguish them less frequently after a while. After repeated perception, we can adapt to the point where we no longer consciously perceive it, and "block it out." Our brain performs several tasks to make images easier to see. Our brain is so accustomed to seeing things upside-down that it eventually adjusts to it. As a result, it is believed that for the first few days, babies see everything upside down. This is because they have not become accustomed to vision. *Our ear,* however, maintains its' upside-down body map which *is the only place on our body where there is an orderly, anatomic arrangement of treatment points that closely follows our body's anatomy and will send out measurable distress signals only when there is an underlying pathology or imbalance.* Any pain that we are feeling and that may be turning our world *upside-down* can be reduced by stimulating these points so that we can begin to feel more balanced and *right side up* again.

# Chapter 7

# The History of Auricular Medicine

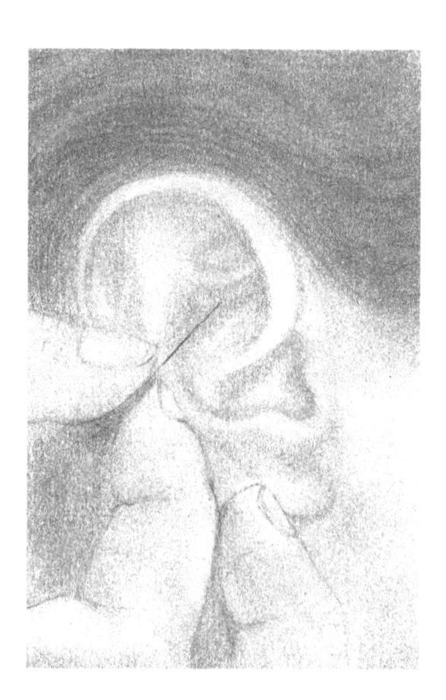

Envision a place where there's no history. We wouldn't have to study it, experience its relics, or explore the people that it would have uncovered. There would be nothing to show us where we've been, or help us to understand how and where to move forward. For many of us, we study history only because we're required too, and what good is learning about history if we don't learn from it? Additionally, it seems like so many things are just left out or presented in ways to emphasize the more comfortable narrative for whomever is teaching it, which is usually the dominant population. For example, when I lived in Hawaii, the emphasis was on Hawaiian history and culture, and since native Hawaiians are the dominant population, history is taught from their perspective, which is different from that of other states.

All of this helps contribute to the situation that we now find ourselves in and if we were observant, we should know what has worked in the past just as we should know what hasn't. Unfortunately, many of us don't even think to consider what was done yesterday let alone thousands of years ago. We see history as an old-fashioned study that's taught by boring people in turtleneck sweaters and beards, confined to academia, and has little relevance to the way we live today. However, I've come to realize that there's much more to it than just memorizing names and dates. History, it seems, politely reminds us to check what's happened in our past before making decisions on our future. It also reveals valuable lessons that have been passed down by our ancestors, and this is certainly the case when it comes to improving our health.

In this chapter, we'll go through the history of how and when the ear was first used to relieve pain and manage other ailments. It tells the story of how the methods used long ago give us continuity to the therapies used today, which in turn provide hope for tomorrow.

Remote tribe in the South Pacific practicing traditions

Traditions of the past give us continuity for today and hope for tomorrow

Dr. Kalfus in Timor-Leste, Maritime Southeast Asia during humanitarian mission 2011

Our ancestors were likely performing a simple form of auricular medicine during the Stone Age and remnants of these primitive treatments have survived in many parts of the world right down to today. Fish bones, bamboo clips, and later various shapes of needles made of metal, replaced primitive sharp stones and bamboo. When stones and arrows were the only tools of war, warriors wounded in battle found that some diseases that affected them for many years were gone. Today, twenty-first century Eskimos living in North America are still using sharpened stones for treating their illness. The Bantus of South Africa scratch certain areas of their skin to lessen the symptoms of many illnesses, while in Brazil there is a tribe whose method of treating illness is to shoot tiny arrows from a blowpipe to specific areas of the skin. The practice of cauterizing a part of the ear with a hot metal instrument has also been reported among certain tribes in Arabia, which is likely a vestige of the acupuncture practiced in ancient Egypt and Saudi Arabia.

Dr. Paul Nogier, the father of modern auricular therapy, developed his theory only after noticing that some patients attending his clinic in France in the 1950s had a small scar from a burn on part of their ear that had been cauterized by a woman in the French countryside for the treatment of sciatic pain.[35] His recognition of this unexpected connection between the ear and the body and his mapping of the ear may be one of the greatest discoveries in medicine.

---

35 Lower back pain that travels down the back of the leg.

## The oldest medical documents known to mankind

Ebers Papyrus from 1500 B.C.E.

The Ebers Papyrus from ancient Egypt (approximately 1500 B.C.E.) is one of the two oldest maintained medical documents[36] and it offers the most complete record of ancient Egyptian medicine. There is a portion of the papyrus (paragraph 856a) that references the *Lower Egypt Den* that would place its origin nearer to the *First dynasty (3000 B.C.E.)*. For this reason, some believe it to be a copy of ancient books of the father of medicine, alchemy, and pharmacy, Thoth (3000 B.C.E.). At any rate, the ailments addressed range from crocodile bites to toenail pains. Included are sections on intestinal disease, asthma, arthritis, migraine headaches, diabetes, pregnancy, birth control, burns, fractures, ear-nose-throat diseases, and dentistry. The circulatory system is described surprisingly accurately,[37] including the role of the heart and the existence of blood vessels, and there

36 The other is the Edwin Smith Papyrus dated about 1600 B.C.E.

37 "Treatise on the Heart."

also is a section on psychiatry, dealing with a condition that would be similar to depression.[38] The Ebers Papyrus also shows a quest for a systematic approach to health and restoring of the natural harmony to the human body. It recognized a connection between the physical and the psychological forces in human health and wellbeing and describes a system of channels and vessels, which closely resembles the Chinese system of channels that is distinct yet in communication with the blood vessels, lymph vessels, and nerves. *It is documented that women in ancient Egypt who did not want any more children had their external ear pricked with a needle or cauterized with heat.* Also in primitive cultures, earrings were believed to keep evil spirits from entering the body through the ears. In addition to their protective power, earrings were thought to have curative effects. *Pierced earrings, particularly when placed in the center of the ear lobe, were recommended to strengthen weak eyes, and gold earrings have adorned the earlobes of men and women since at least 3000 B.C.E.;* those set with emeralds were considered particularly effective. For those seeking to cure headaches, gold ear jewelry was adorned.

---

38 "Book of Hearts."

## The father of Greek medicine and the Roman Empire

Gaius Julias Caesar of the Roman Empire

Hippocrates (460–375 B.C.E.), the father of Greek medicine, reported that doctors made small openings in the veins situated behind the *ear* to facilitate fertility and reduce impotency problems. Cutting of veins situated behind the ear was also used to treat leg pain. The Greek physician Galen (AD129–216) introduced Hippocratic medicine to the *Roman Empire* in the second century CE, and commented on the *healing value of lacerating and piercing at the outer ear.* Galen, known as the greatest physician of the Roman Empire, viewed health as the balanced, harmonious, optimal functioning of all the organs and systems of the body. He was an *expert on the pulse, and he is considered the originator of pulse diagnosis.* Today, nearly two thousand years later, European auricular medicine and

Traditional Chinese Medicine both utilize a form of pulse diagnosis that involves palpation of the radial pulse on the wrist. The Nogier pulse (or Vascular Autonomic Signal) looks for changes in pulse strength in response to a stimulus placed near or on the external ear, whereas Oriental doctors monitor the steady-state qualities of the resting pulse and look for changes in pulse rate and rhythm.

## The loss of medical knowledge

After the fall of the Roman Empire (around AD 476), much of the medical knowledge from the Roman and Greek civilizations was lost, consequently the quality of medical practitioners was poor. The Roman Catholic Church now dominated politics, lifestyles, beliefs, and thoughts. It wasn't an environment conducive to creativity or science and thinking differently from church teachings had grave consequences. Research, development, and observation gave way to an authoritarian system, which undermined scientific thinking. Ideas rarely had the chance to travel because communication between territories run by feudal lords was poor and perilous. *The only places that managed to continue learning and studying science were the monasteries.* In many places, monks were the only people who knew how to read and write. Fortunately, Muslim cities in the Middle East had translated the medical records of Egyptian, Greek, and Roman medicine and kept them in their centers of learning. Thus, this *knowledge was best preserved in ancient Persia and the Arabian world.* Included in these Persian records were specific references to medical treatments for sciatic (lower back) pains and sexual-related ailments. The remedy was the cauterization[39] of the outer **ear**.

[39] That is to treat by burning with a heated instrument.

## The beginning of hospitality

The Dark Ages

In the Dark Ages (AD 500–1500), European Medieval hospitals had a slightly different meaning compared to what we understand today. For those in desperate needs, the Christian thing to do was provide *hospitality*, i.e., food and shelter, and medical care if necessary. The Hotel-Dieu, founded in Lyons in 542 AD, is the oldest hospital in France.[40] Throughout Europe, monasteries provided medical care and spiritual guidance in *hospitiums*, where the monks provided the expert medical care and the lay people were helping them. During medieval Europe, barbers, not doctors, were in charge of surgery. Also during the thirteenth century, wine was noticed to be an effective antiseptic useful for washing out wounds and preventing infection. The Middle Ages gave way to the Renaissance and advances in medical practice accelerated dramatically.

---

[40] Coincidentally, Dr. Nogier is from Lyon.

## The Renaissance

During the *Renaissance*, which began in the fourteenth century, there were occasional clinical reports in Europe that described the use of *ear cauterizations to relieve leg pain.* Europe starting trading with nations from all over the world, which was good for wealth and many people's standards of living, but it also exposed them to pathogens from faraway lands. Diseases from Spanish explorers came to the New World and circulated in the Americas where Native Americans had no immunity against such diseases. The *Dutch East India Company actively engaged in trade with China from the 1600s to the 1800s, and its merchants brought Chinese acupuncture*

*practices back to Europe.* Doctors working with the company had become impressed by the effectiveness of needles and cauterization of the outer *ear*, or by cutting the veins behind the ears for relieving conditions such as sciatic pains and arthritis of the hip. *In 1637*, the Portugese physician Zacatus Lusitanus, *for the first time in Europe, described the treatment of sciatic pain by cauterization of the ear* after bloodletting had failed. The Italian surgeon Antonio Maria Valsalva (1666–1723), who made the first modern anatomical description of the ear in 1717, published the *Aura Humanus Tractatus,* where he describes the treatment of toothache by scarification of the antitragus of the outer *ear.* In 1810, Professor Ignazio Colla of Parma, Italy, reported the observation of a man stung by a bee in the antihelix of the ear,[41] which resulted in dramatic relief of pain in the legs, and in the same year, Dr. Cecconi of Italy performed cauterization to help treat sciatic pain. In 1850, a French medical journal reported thirteen different cases of sciatic pain that had been treated by cauterization with a hot instrument applied to the *ear.* Only one of the patients did not improve completely. But *it was not until a century later that famed neurologist Dr. Paul Nogier of Lyon, France, rediscovered this type of treatment.*

## The first publication of acupuncture in the United States

In China, widespread use of acupuncture diminished in the 1800s when it became dominated by imperialist powers from Europe. In 1822, the minister of health for the Chinese emperor commanded all hospitals to stop practicing acupuncture. The erosion of faith in traditional Oriental medicine followed the defeat of the Chinese military in the Opium Wars of the 1840s. Because traders from Europe had more powerful weapons than the Chinese, it came to be believed that Western doctors had more powerful medicine.

*In the United States, the first publication on the use of acupuncture was in 1826 by Benjamin Franklin Bache, the great grandson of Benjamin Franklin.*[42] *In the Civil War Journal and Surgical Guide,*

---

41 The curved prominent part of the ear just inside the outermost rim.

42 Although articles first appeared in medical journals in 1822 and Bache tested

*acupuncture was discovered as a way to stop hemorrhaging.* In 1906, American and British missionaries founded the Peking Union Medical College in China, which in 1915 was purchased and upgraded by the Rockefeller Foundation.[43] It was modeled after one of the finest medical colleges in the world, Johns Hopkins University in the USA. Its purpose was to improve the medical and hospital conditions in China. Fortunately, the discoveries of Dr. Nogier's ear charts arrived in China in 1958. Although ear acupuncture was used throughout China prior to their learning of Nogier's therapy on the topic, *the correspondence of the ear to specific body regions was not shown on ear acupuncture charts prior to 1958.* This resulted in a massive movement to study and apply ear acupuncture across the nation. In the classic Chinese text, *The Mystical Gate: Treatise on Meridians and Vessels,* the Chinese thought was, "the ear is connected to every part of the body because of the ceaseless circulation of energy and blood through these meridians and vessels. The ear joins with the body to form the unified, inseparable whole."

---

the efficacy of acupuncture needles in 1825.

43 Internal Memorandum, The Rockefeller Foundation, March 7, 1915, Rockefeller Archive Center (RAC), RG 2, Family Records, Series O, Box 11, Folder 92.

## The discoveries of Paul Nogier

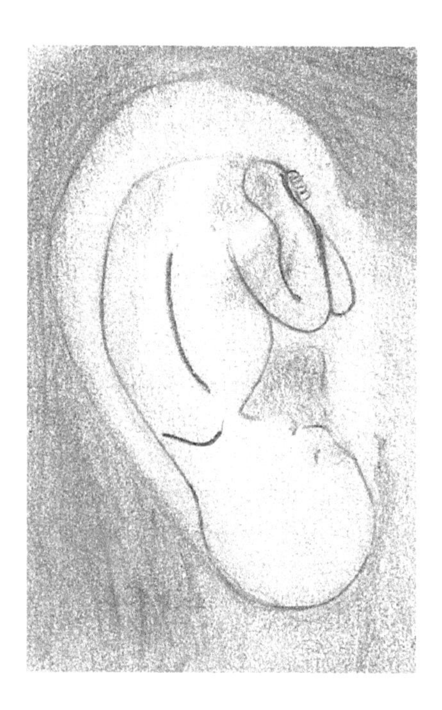

In 1957, Dr. Paul Nogier first presented his observations that a person's entire body is represented on the ear in the same anatomic arrangement as our body is arranged, but in an upside down position. This meant that the top of the ear represented our feet while the bottom of the ear represented our head.

He developed this theory after noticing that some patients attending his clinic had a small scar from a burn on a part of their ear. On inquiring into this, he was told that a very small area of their ear had been cauterized by a certain Madame Barrin for the treatment of sciatic pain—a treatment that proved very rapid and effective. Later, he mapped out the entire ear in an arrangement that was the same as the human body. Points on the body, for example the knee, correspond precisely with the knee point on the miniaturized human body that is represented in the ear.

Dr. Nogier found that the ear was useful for evaluating the balance between the left and right sides of the brain as well as for the detection and treatment of various imbalances throughout the body. He was also able to locate any blockages to treatment, such as emotional disorders or scar tissue from a previous injury by looking for changes in his patients pulse while he was pressing on the ear. Dr. Nogier discovered a distinct change in the strength of a person's pulse when stimulating certain points on the ear. He called this pulse response the Vascular Autonomic Signal (VAS) and it's an important principle of auricular medicine. Interestingly, the Greek physician Galen of the Roman Empire also used the pulse for diagnosis. As Dr. Rene Bourdiol[44] eloquently stated in the Elements of Auriculotherapy,[45] "*render unto Caesar what is Caesar's and auriculotherapy to NOGIER.*"

## From Nogier to today

Dr. Nogier's research on the ear was published in a German Acupuncture journal in 1957. This journal had an international circulation, and it wasn't long before Japanese acupuncturists became familiar with Nogier's mapping of the ear. This discovery soon spread to China and led to a massive study by the Nanking Army Ear Acupuncture research Team. This Chinese medical group verified the clinical effectiveness of the Nogier approach and assessed the conditions of over 2000 clinical patients, recording which ear points corresponded to specific diseases. The outcome of that research was very positive and resulted in the utilization of this therapy by the "Barefoot Doctors" who roamed the countryside during the Cultural Revolution. China published an Ear Chart remarkably similar to that of Dr. Nogier in 1958, and it remains unchanged today.

In 1972, Dr. Nogier made his first visit to the United States and met with Dr. Jim Shores in Atlanta, Georgia. He introduced to America his research on the ear and the different ways to diagnose and treat ailments, which included the use of small electrical currents on the ear. Also in 1972, President Nixon visited China and his team witnessed acupuncture firsthand and spoke highly of it upon returning to the United States. Within a few years, research provided us with a dramatic breakthrough. In 1980, a study at UCLA by Richard Kroening and Terry Oleson *verified the scientific accuracy*

[44] A colleague of Paul Nogier.

[45] Maisonneuve 1982.

*of Auricular diagnosis* that was being claimed by both French and Chinese mapping of the body on the external ear. *The consistency between the established medical diagnosis and the auricular diagnosis was 75.2 percent.* More recently, modern images of the brain have clearly demonstrated that treatment of the ear contributes to changes in brain activity that correlate to a reduction in pain associated with a wide variety of health issues and concerns.

In summary, medical documents that are over three thousand five hundred years old described in detail how the ear was used to treat numerous health conditions to include fertility, improving eyesight and lower back pain. Monitoring the pulse was also recognized as a diagnostic tool since the time of the Roman Empire. More recently, scientific studies have verified the consistency between established medical diagnosis and auricular diagnosis. History, it seems, politely reminds us that if we eliminate the impossible, whatever remains, however implausible, is likely to be the truth.

# Chapter 8

## A Truth Spoken Before Its Time Is Dangerous

In pain management, doctors and researchers throughout the world have dedicated themselves to finding ways to safely and effectively relieve pain. In order to determine if a treatment is actually doing what its designed to do, scientists will measure the differential effect observed between the true therapy as compared to a placebo or fake treatment using a scientific and methodological design. But when all is said and done, summarizing all of the available information can result in a wide range of truths and fallacies, depending on who is interpreting the findings.

## The parable of the blind men and an elephant

The parable of the blind men and an elephant is a good way to describe this process. Here, a group of men in the dark touch an elephant to learn what it's like. Each one feels a different part, but only one part, such as the side or the tusk. They then compare notes and learn that they are in complete disagreement. While one man's subjective experience is true, it may not be the totality of truth. The most comprehensive way to describe "the elephant" is to look at the entire body of evidence, including any inconsistencies in data, along with the procedures that have stood the test of time and clinical outcomes in order to see a clearer reflection of reality. And if truth is what stands the test of experience and it cannot be long hidden, the key then is to discover it. For once discovered, it will be easy to understand. Furthermore, since a comprehensive review of modern research has demonstrated the mechanisms of auricular medicine and confirmed its efficacy on a multitude of disorders, it no longer seems possible to doubt its truth.

# Chapter 9

## What Makes Chronic Pain Unique?

Everyone has experienced physical pain. For many of us, it's only experienced for a brief period of time, such as when we stub a toe or have an occasional headache. But for others, pain can continue long after the usual time of recovery. If it continues and becomes recurrent, it can assume a central role in our life.

While all pain is real, *chronic pain is much more driven by our emotions* (i.e., fear, guilt, anger, loneliness, and helplessness) *and thoughts.* The process of thinking and then feeling in response to our thoughts has an influence on the experience of pain. This is why our words (what we say and what we hear), our thoughts, and even the way we breathe are so important in pain management (see Chapter 2).

In my experience, the main characteristics of chronic pain are as follows:

- It lasts for more than twelve weeks.
- It often becomes the center of a person's life.
- It often persists despite traditional treatments such as drugs and exercise.

  In fact, pain can actually increase with extended drug use—a condition known as drug-induced hyperalgesia.

- The extent of the pain and any disability is often in excess of what we would expect based on our examinations and tests.
- It often comes with feelings of depression.
- The daily use of medication frequently leads to drug dependency.

In summary, *it's our emotions and thoughts that drive the experience of chronic pain and make it different from acute pain.* Unlike any other treatments available, auricular medicine is unique in that it provides us with a personalized remedy that's based on the needs of our body. And our ear gives us exclusive access to our nervous system, which is where we interpret and express pain based upon the interaction of physical, psychological, and emotional factors.

Our body is telling us what it needs, but we need to listen. We need to recognize that *there is a unique healing language that has been imprinted in each of us.* Thankfully, we now have the technology and the training to understand this language and to give our body exactly what it is asking for.

# Chapter 10

## The Virtue of a Touch

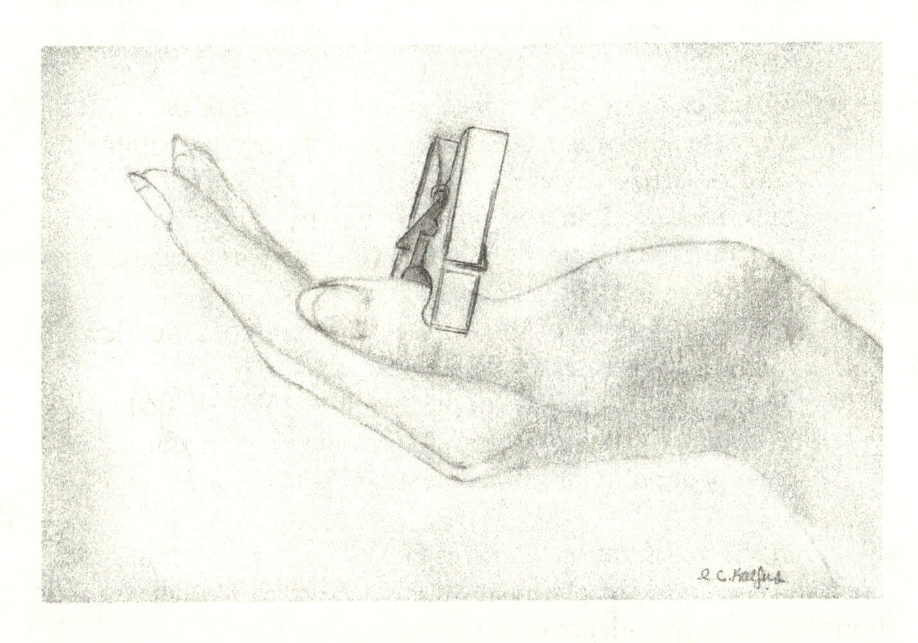

All too often, we underestimate the virtue of a touch, a smile, a kind word, and the smallest act of caring or a listening ear. And when we do listen, it's often with the intent to reply rather than to understand. Today, I am asking those of you who have eyes to see and those of you who have ears to hear; to lend an ear to the thousands of years of historical medical knowledge as it's integrated with modern science and validated by research throughout the world. Something as simple as a touch, or as sophisticated as a light flickering on and off at very select frequencies can be used to detect imbalances in our body and accelerate our natural healing process.

When I first began to study with Dr. Nogier and Dr. Oleson, they both encouraged me to keep an open mind and to look at the science, the years of research, and the clinical data that supports its use. Although this approach to healthcare is different and far less invasive from what we commonly use here in the United States, the results, the relative ease of treatment, and its safety are just too impressive to overlook.

## What we can see when we open our eyes to our ears

A skilled practitioner is able to unveil a wealth of information about someone's health simply by observing the following on the ear:

- Visual changes in the appearance of the skin of the ear (i.e., a crease in the ear lobe indicates a higher probability of cardiovascular disease).
- Subtle changes in the pulse when the ear is exposed to different frequencies of light (may indicate an imbalance in a particular area of the body).
- Electrical activity of the ear (i.e., an increase in the flow of electricity indicates an imbalance).
- Sensitivity to touch at specific points on the ear that elicits a "grimace" response or an expression of pain (indicates a treatment point for immediate pain relief).

All of this information, which is being broadcast by your entire body and is uniquely displayed on your ear, is here to benefit you and is more than a coincidence or a mere whisper of hope.

Modern technology such as functional Magnetic Resonance Imaging (fMRI), which measures brain activity, has clearly shown that stimulation of the skin causes a measurable reduction in activity in the pain centers of the brain. Dr. David Alimi of France[46] demonstrated that stimulation of specific areas of the outer ear led to selective changes in fMRI responses in the brain. Furthermore, Dr. Z.H. Cho, the neuroscientist and inventor of the first PET-MRI imaging device,

[46] Alimi, 2000; Alimi et al., 2002.

shared with me the images he had taken of the brain that compared brain activity before and after a person's skin was stimulated with tiny acupuncture needles. As we can clearly see in the images below, the activity in the brains pain centers is greatly reduced.

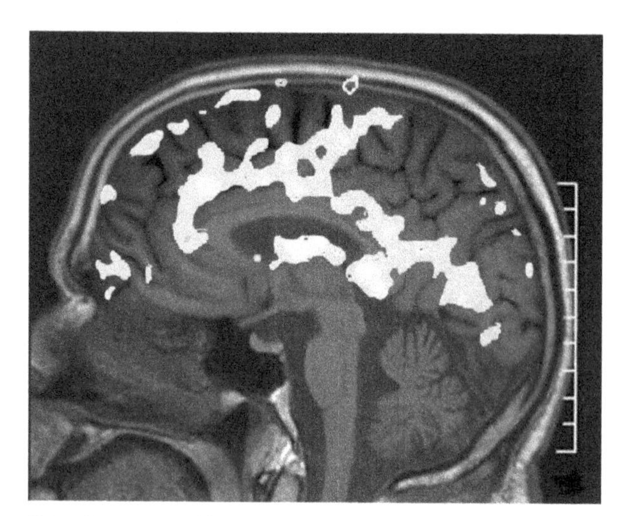

fMRI image of brain courtesy of Dr. Z. H. Cho
(Brain activity to provoked pain /**before**-acupuncture)

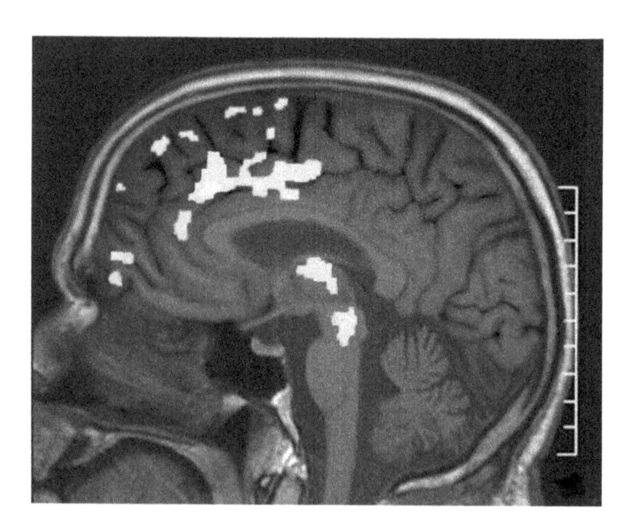

fMRI image of brain courtesy of Dr. Z. H. Cho
(Brain activity to provoked pain /**after**-acupuncture)
Note: The lighter areas are smaller (less excitation) following acupuncture indicating a reduction in pain.

For those of you who still don't believe that there's a correspondence between our ears and our body, here's an easy way to demonstrate it. By simply evoking pain in any area of the body,[47] a corresponding point in the ear will become sensitive to the touch. This point will return to normal and no longer be responsive to touch when the corresponding area of the body[48] is no longer in pain.[49] Chapter 12 will talk about this in more detail. Also, by treating the area of the ear that is painful to the touch, which corresponds to a painful area on your body, the pain you were feeling will quickly disappear.

Long before the first pharmaceutical drug was ever patented or sold, ear treatments were being used safely and effectively. *Today in continental Europe, tens of thousands of doctors use auricular therapy on a daily basis, and they've been doing so for over forty years.* It's time for us here in America to benefit from Dr. Nogier's discoveries. In addition to its proven ability to relieve pain and help regulate all of our body systems, treatments of the outer ear will also help with a wide variety of health issues and concerns, often after all other medical interventions have failed; and all by the virtue of a touch.

---

47 That is pinching your thumb with a clothes pin.

48 In this example your thumb.

49 That is the clothes pin is removed from your thumb and your thumb returns to normal.

## Pain is both a personal and public health challenge

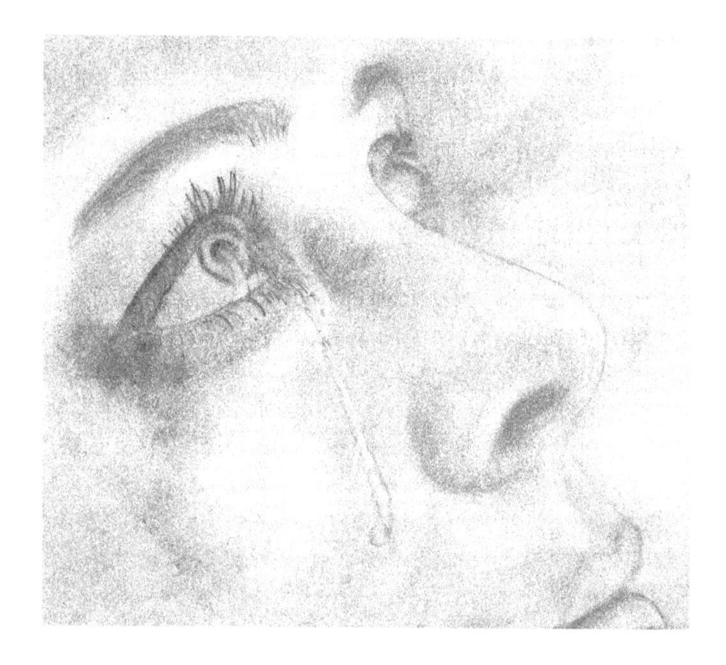

According to the American Academy of Pain Medicine, chronic pain affects over one hundred million Americans annually and costs our society over $600 billion annually, which includes $300 billion in healthcare costs and over $300 billion in lost wages due to missing work. The Center of Disease Control and Prevention (CDC) and the National Center for Health Statistics (NCHS) report that most people in chronic pain have multiple sites of pain which include low back pain (28.1 percent), knee pain (19.5 percent), severe headache or migraine (16.1 percent), neck pain (15.1 percent), shoulder pain (9.0 percent), finger pain (7.6 percent), and hip pain (7.1 percent). The prevalence of chronic pain in the United States has risen to over 40 percent of the adult population (excluding acute pain and children) and in England, for example, it is three times more common now then it was forty years ago according to the U.K. Department of Health. It's clear that pain is both a public and personal health challenge and as such, we need to implement treatments that are safe, proven, and cost effective.

## The new normal

The battle you are going through with chronic pain is fueled by your mind, which gives it importance. *Pain has become the new normal* for so many of you because your mind sees it as such, and in a body that has lost the ability to regulate its own systems, *auricular therapy can stimulate an inborn regulatory system in a manner that induces self-correction mechanisms.* Your regulatory system releases and balances the circulatory levels of the same natural proteins, such as pain-relieving endorphins, that immune cells use to communicate with each other, and every other system in your body. This form of communication has been demonstrated by several scientists.[50]

Sometimes, the chronic pain messages that you've lived with for so long just need to be "unlearned" and replaced with pain-free messages that will allow you to return to a healthy, balanced, and more comfortable condition. As an example, when you go on a walk, you usually prefer to take the path that you're most familiar with and that's used most often, like a sidewalk or a well-worn trail, rather than going off the trail, which is usually more difficult, takes a lot more effort and is unknown. Your mind works in a similar way and prefers to follow patterns that it's familiar with. *Auricular therapy re-familiarizes your brain and your body with these pain-free messages.*

---

50 Becker RO, Marino A, 1982; Pert CB, Dreher HE, Ruff MR, 1998; Weigent DA, Blalock JE, 1995.

Chapter 11

# Common Myths about Auricular Therapy

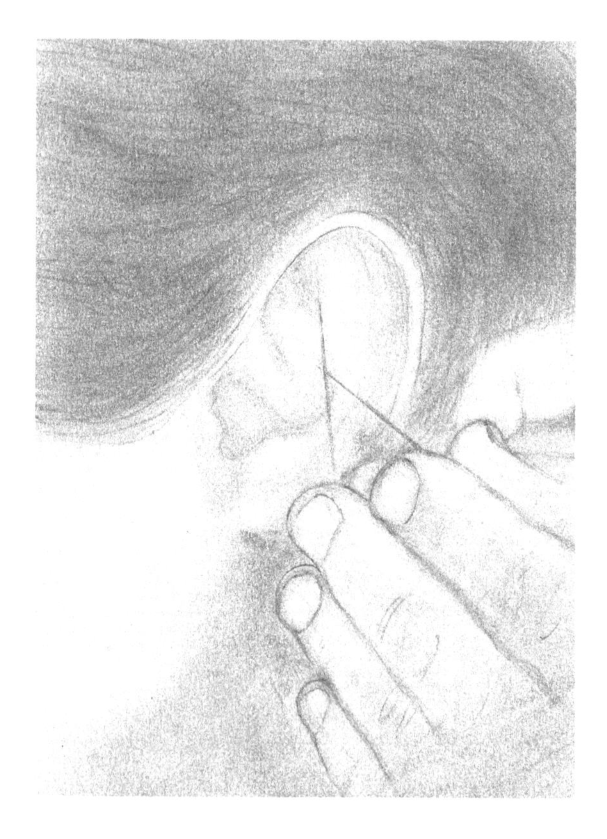

One of the definitions of a myth is a false idea that looks like a truth. Some myths come from hearing what others around us say and some come from our own misunderstandings. Whatever the source, it's always important to question things that sound like truths.

A helpful way to expose the myth that auricular medicine points[51] are the same as traditional acupuncture points found anywhere in the body is to know how they differ. Here are some of the differences:

- An acupuncture point never moves anatomically. It's always fixed in one place. Whoever the person, and whatever the disease or time of the examination, it's always located at an identical area with the same anatomical landmarks.
- An acupuncture point is always detectable. It always demonstrates electrical activity whether a person is healthy or not. Even a few days after a person has passed away, their body has points that can be detected electrically.[52]
- The *auricular medicine point,* in contrast to the acupuncture point, is neither fixed in one place nor detectable at all times. *In order for it to "exist" at all on the ear, there must be an underlying imbalance or pathology in the body.*[53] It serves as a beacon and provides us with invaluable information on both the location of the problem and how to best treat it. *The auricular medicine point* also *moves*[54] and its location is dependent upon whether the problem is acute, chronic, or degenerative.
- An acupuncture point only affects the same side of the body being treated and stimulating it always acts on the same part of what Traditional Chinese Medicine refers to as a meridian (energy highway), which has acupuncture points along its path.
- The *auricular medicine point* does not always act on the same side of the body. Its action *may be direct, or crossed to the opposite side of the body.*[55]
- An acupuncture point always responds in the same way. Whatever the type of stimulation (message or needle), the result is always the same.

---

51 Treatment points found on the ear.

52 Dr. J. E. H. Niboyet and Professor J. Bossy.

53 That is it is only detectable when there is a underlying problem.

54 In contrast to being fixed in one place.

55 That is stimulating the right ear may help the left side of the body.

- The *auricular medicine point does not always respond in the same way. There are two types of auricular medicine points and each responds to different treatments.* A point sensitive to touch will respond well to a tiny needle, while it may not respond at all to light therapy. Likewise, an ear point that's not sensitive to the touch and is found only by electrical detection (to be discussed later) will respond to light and may not respond to treatment with a needle.
- The *auricular medicine point exhibits interactions and true relationships with other points.* Points along the outer rim of the ear will "turn off" and deactivate more inward points while inward points will "turn on" and activate more peripheral or outward points. On the contrary, acupuncture points are always active and not turned on or turned off based on the treatment of other points around them.
- *Auricular medicine uses bipolar electrical detection* (to be discussed later) *to precisely locate the electrically active treatment points on the ear.* The skin of the ear is relatively thin and is uniquely sensitive to electrical detection. Bipolar detection cannot be used to detect body acupuncture points because body points are always electrically active, regardless of health. Ear points only become electrically active when there's an underlying problem in the body.
- *Auricular medicine points that are electrically active* can be treated with light, which will cause a momentary change in pulse strength.
- *Auricular medicine points provide unparalleled access to our nervous and immune systems.* The only area on our entire body where we have superficial access to the vagus nerve is our ear. *The vagus nerve helps to "calm us down" and relax us by lowering our heart rate, blood pressure, and our breathing as well as relieving anxiety and depression.* It oversees a vast range of crucial functions and *communicates with every organ in our body.*

As you can see, there's quite a difference between the auricular medicine points that are found exclusively on our ear and the Traditional Chinese acupuncture points that are found throughout our body.

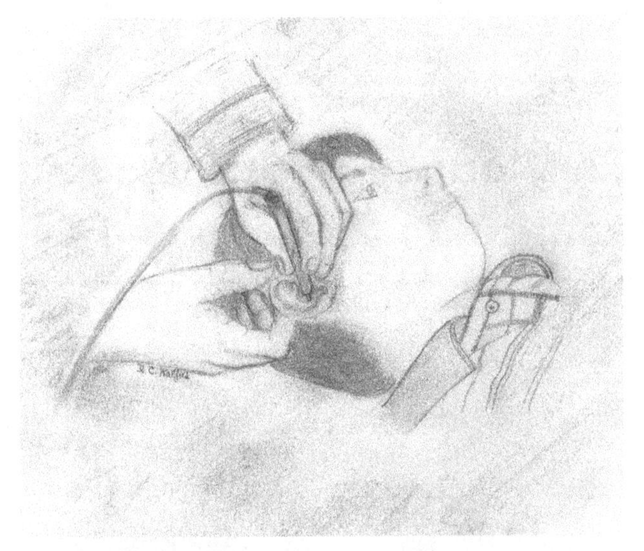

Auricular medicine points move and can be located with bipolar detection

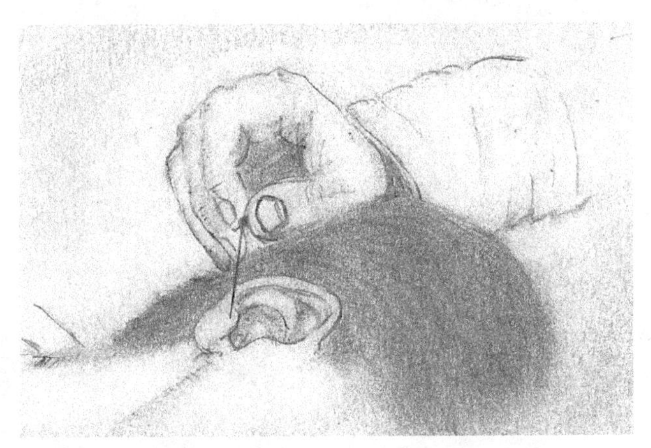

Traditional acupuncture points are fixed in the same location and never move

# Chapter 12

# Ear Points Are Strange and Beautiful to Well Opened Eyes

If there was ever a time when I had just one chance to show other doctors how effective auricular therapy was, it was in Balboa, California, in 2015. While visiting a colleague, I was asked if I would like to see one of his patients who he had been treating for the past few years. She had been seen by a number of specialists and was taking several medications to help with overall weakness, numbness, and pain throughout her body. She was diagnosed with a condition known as peripheral neuropathy, which results when something is wrong with the nerves that carry messages to and from the brain and spinal cord, and to the rest of the body. It occurs in only about two out of every hundred people in the general population but is more common in people with underlying conditions such as diabetes, vitamin deficiency (B12), and alcohol abuse to name a few. Since it was known that I had personally trained with the renowned Dr. Nogier, the doctors in this hospital wanted to see if there was any truth to using the ear as a way to treat pain and stimulate healing. The patient was more than willing to try something different and she herself had heard of Dr. Nogier.

As with so many chronic pain patients, when I first saw her, she had all but lost hope of ever feeling better. Without any preconceived notions on where to look, and under the watchful eyes of colleagues, I scanned her ears with a pulsed light and quickly found a few areas' that responded to frequencies,[56] which under healthy conditions would not cause any response. For precision, I located the electrically active areas with a bipolar electrode and then treated the points on the ear with both a pulsed light and some tiny needles that look like very small earrings. These "semi-permanent needles" from France remained in her ears for a few days and then fell out on their own. At her one-week follow-up, to the surprise of the staff, she said she had felt better than she had in years, had slept well all week, and had not taken any medication for pain since her visit. After she gave everyone a big hug, I replaced some of the French needles that had naturally fallen out, bathed the needles in pulsed light, and she went happily on her way. Her pain is now very manageable and she no longer relies on pain medication. Her tongue, which at first appeared shiny-smooth and slightly red, also returned to normal indicating that she's better absorbing her Vitamin B12, a deficiency that may have contributed to her pain. I was able to help her simply by paying attention to the information that her ear was providing me about her body. As for my colleagues, they frequently refer patients to me, and it's my hope

56 As noted by an increase in pulse strength.

that they will soon provide this treatment as an option or adjunct to drugs, surgery, and exercise.

## Two types of ear points

One of the most common misunderstandings in auricular therapy is that our ear will respond to any kind of stimulation. Whether it's a gentle light, electrical treatment, or tiny needles in the ear, the outcome will be the same; an ear point is an ear point. Unfortunately, this is wrong. Science has shown us that there are actually two distinct types of points and each one responds only to a particular type of treatment. So if we get this wrong, the therapy won't work, and if we just guess, the results won't be consistent.

This chapter will explain the differences between the two types of ear points, and I'll share with you the best ways to treat them. You will also see just why the outer ear is so unique and a true game changer when it comes to the alleviation of pain and speeding up the healing process. At first, the ear points may appear strange, as they did to me. But in time, they become very beautiful to well opened eyes. I refer to them as beautiful because they tell us exactly what our body needs.

There are only *two kinds of ear points and each is very distinct from the other*: First, there are those points that connect directly to your nervous system and are painful to the touch only when there is a corresponding problem.[57] Second, there are those points that can only be located by electrical detection and do not hurt when they are touched. These points have a lower electrical resistance to the flow of electricity, and they also have very specific microscopic features.

---

57 That is if your shoulder hurts, the corresponding shoulder point on the ear will also hurt when touched.

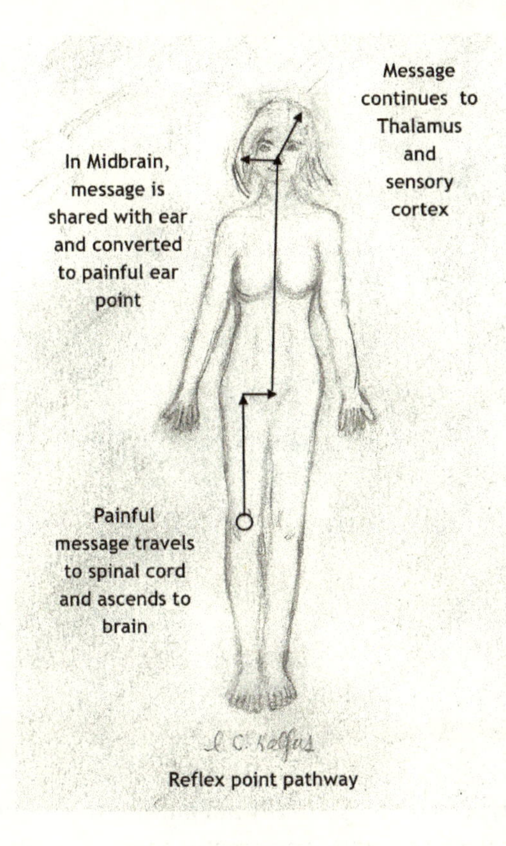


Reflex point pathway

## A Reflex point

*The points that are connected directly to your nervous system are called reflex points.* They are located by touching the outer ear with slight pressure and they become uncomfortable or painful to the touch only when the organ or tissue that corresponds to that point is unhealthy. Oftentimes, these points are easy to find and can be quickly treated with message, pressure or a tiny needle. *The response to treatment will be immediate.* Reflex points make up only 10–15 percent of all the ear points.

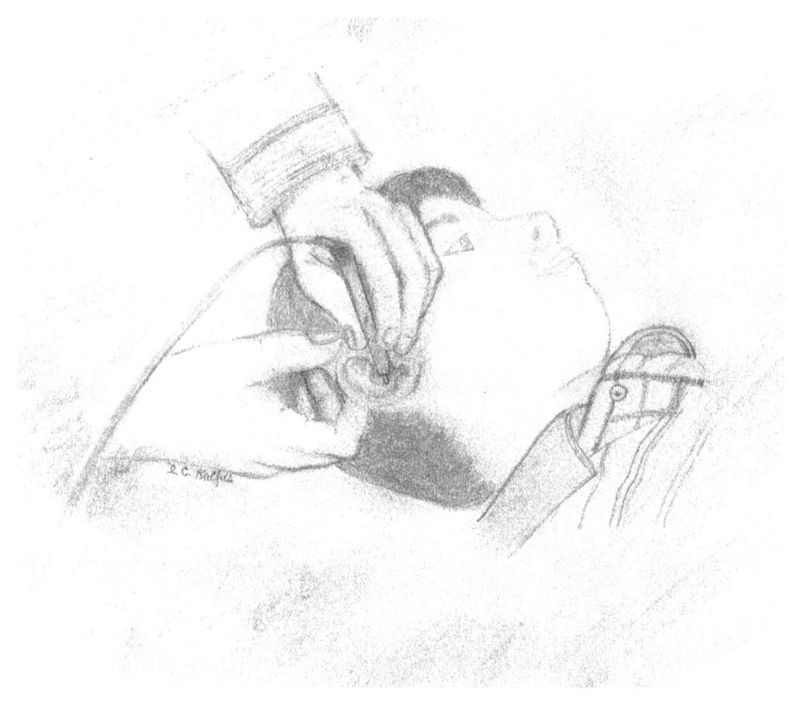

An electrically active ear point

*The second type of point is usually not painful when touched and is known as an Electrical point.* Many practitioners also refer to these as neurohumoral or neurovascular points. I prefer to call them electrical points because they emit electricity. *These points are by far the most numerous and can only be located by precise electrical detection.* Their value in treatment can be quickly verified by the monitoring the level of electricity that's being emitted by the ear and by seeing if the patient's pulse changes when a point on the ear is stimulated. These points have very specific structures within and below the skin, and will be discussed later. *Using either pulses of light*[58] *or minute electrical stimulation in the millionths of an amp is the best way to treat these points. Both methods are completely painless, often needing only a few seconds of therapy.* Technology has enabled practitioners to precisely locate and treat electrically active points, which are only 0.2 mm in size.[59] This means that if we are off by 0.5 mm, we will miss the

58 Ear points are very sensitive light.

59 Eight thousandths of an inch.

point. *When an area in our body is not healthy, our ear will emit more electricity.* State-of-the-art, FDA approved medical devices can detect electrical changes, which tell us that a point has become "active." Not all points on the ear are active, and some are more active than others. Only the most active points should be treated, and they are only active when there is an underlying problem with the corresponding organ or tissue.

The only electrical detection devices that are scientifically proven to work on the ear are those that compare the electrical activity at the center of the point being evaluated to the electrical activity of a small area surrounding the point. The illustration below shows these two areas.

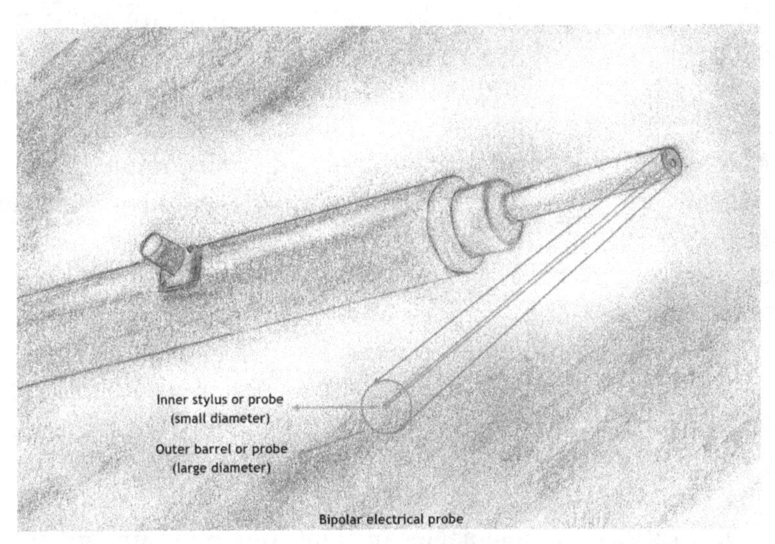


Electrical detection instrument that touches the skin

This difference in electrical activity between the inner stylus and the outer barrel of the probe is what scientists call the "Mega-Ohm differential," and it determines whether or not a point is emitting electricity and needs treatment. Note that the diameter of the outer barrel of this instrument is only 3.0 mm,[60] and the inner stylus is only 0.67 mm[61] making detection of an active point a very precise

[60] About 1/10th of an inch.

[61] About 1/32nd of an inch.

science. This procedure is also very comfortable and it takes just a few minutes to scan the entire ear.

The doctor has the option to treat an electrically active ear point with the same probe by selecting the proper electrical frequency, strength, and length of treatment. This instrument is actually two separate instruments in one. When it's set in the detection mode, it detects for electrical activity. When it's set in the treatment mode, it will deliver a very small current of electricity through the smaller center probe. Both detection and treatment are very gentle.

## Our skin receives and interprets light

We know that our skin and our nervous system are very similar in make up. They are both derived from the same group of cells when we first develop. I think of our skin as being like a telescope; it collects light from our surroundings and uses the information that's carried in the light to stimulate our nervous and immune systems. Just as we[62] use the minerals, plants, animals, liquids (water), and gases (oxygen) from our surroundings for nutrition, we also take in the electromagnetic waves, or light to help regulate our nervous system. In other words, our skin has photo-perceptive properties, which means it's capable of receiving and perceiving (interpreting) light. Photoperception is the sensitivity of our skin to light and its ability to transmit information coded in the light to our brain and spinal cord.

When either the skin on our body or the skin on our ear is exposed to different frequencies of light, our body unconsciously responds by contracting or relaxing the small muscles that line the walls of our arteries. A doctor who feels a change in our pulse strength can observe this, and this happens as a result of the photoperceptive properties of our skin. In a healthy person, these frequencies of light are recognized as normal and only specific areas on our body and on our ears, which have been meticulously mapped out, will respond to only a single frequency of light. All other frequencies, under normal conditions, will not elicit any changes in our pulse. *Drs. Paul and*

[62] Along with other forms of life.

*Raphael Nogier established, in 1977, eight frequency zones on the ear that match perfectly to the eight frequency zones on the skin of the body and each will react normally*[63] *only to a single base frequency.* However, *when we are in pain or are sick or injured, additional frequencies of light*[64] *will also elicit a change in our pulse strength*[65] *when the light is exposed onto the injured area(s) of our body and the corresponding area(s) on our ear(s).*

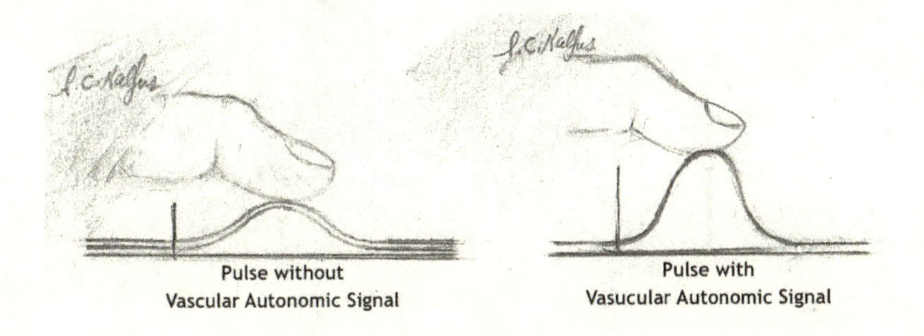


When we are in good health, our skin, which is endowed with photoperception, is *active* all throughout our body, *except for on our ears.* However, *in the event of disease, our ear points "turn on" and will show active photoperception in order to return our bodies back to a healthy condition.* The skills of detecting subtle changes in our pulse as well as knowing which light frequencies are normal and which ones aren't in any given part of the body are important ones for a practitioner to have. There are instruments that emit the light frequencies discovered by Dr. Nogier and allow us to detect imbalances, which contribute to pain and illness. As an example, if you had a sore shoulder, I would shine different frequencies of light on it and note which ones would cause a change in your pulse strength. Under healthy conditions, your shoulder will only respond to a single frequency of light. Under unhealthy conditions, your shoulder will respond to more than one frequency of light. I would then expose your ear(s) to the same frequencies of light that I used on your shoulder, and I would look for a point on your ear(s) where

63 A unconscious change in pulse strength.

64 Which become the "signature" of the problem.

65 See illustrations below for pulse.

I find the same pulse response as seen on your shoulder. When I find that point, I would treat it with those same frequencies of light. And as the French say, "Viola" (pronounced vwa-la), which suggests the appearance of something special. The pain will go away! The treatment is that easy.

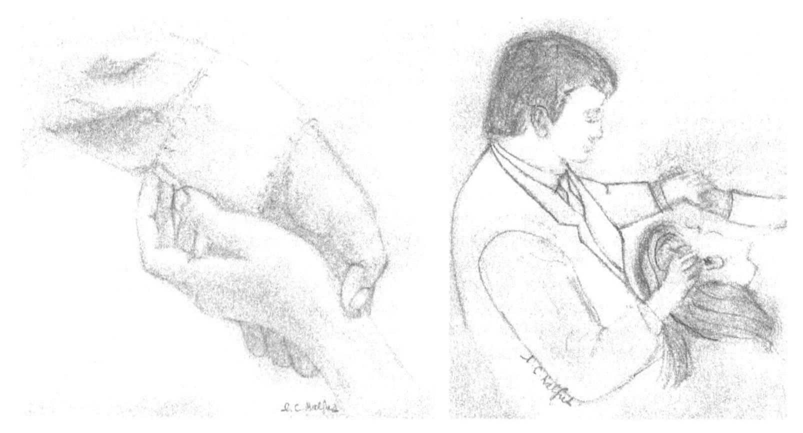

Monitoring the pulse for subtle changes in pulse strength

## Our ear is uniquely wired to the entire body

As mentioned above, there are two distinct types of ear points. The first type is called a reflex point. When a particular part of our body is in pain, there will be a corresponding point of the ear that, when touched, will also be painful. When located, we will often say "ouch" or make a slight grimace, which indicates that this point needs to be treated. If we provoke pain on any part of our body, like our foot or shoulder, we will find that there will be a corresponding pain in a particular part of our ear when touched. Studies were even done on dogs where, for example, a painful response was provoked on the knee and a researcher looked back into the ear to locate a point that also was painful. Although dogs do not verbalize, they made sounds that indicated that the ear point, that previously provoked no response, was now painful. The dog would "woof" it out. The important point here is to know that *the proper time to place a tiny needle in a specific ear point is when there is a painful response to touch. If a point does not provoke a grimace, it's not a valid point for treatment.* Also, a painful point will not respond well to laser or frequency (light or electrical) therapy. If you hurt your shoulder, we will find the painful point on your ear that corresponds to your shoulder, and treat it with a tiny needle. Immediately, the pain will be resolved. Interestingly, with shoulder pain, the point on the ear that represents the shoulder will be painful to the touch 70 percent of the time. In 30 percent of cases, there will not be a painful shoulder point, and this is usually due to patients taking painkillers, which make point location difficult. It's that simple. *Treating pain is simply treating a painful point with a mechanical stimulus such as a tiny needle (or a message or applied pressure). Treating a painful point with frequency stimulation such as light or electrical stimulation will not resolve the pain.* For example, let's say that I step on your foot. The pain message travels up the spinal cord to your brain. Before it reaches the area of your brain known as the thalamus, the message branches off and is also sent to your outer ear. It's truly amazing that *messages traveling through our spinal cord and conveying information about our entire body are being shared with our ear.* There is a simple conversion of the message traveling to the ear[66] that will make one point on our ear painful to the touch (see diagrams below). *Your brain and your ear are hooked up to the same information system.* Think of having a group of phones all connected to a central operator. Years ago, for example, when two people were having a telephone conversation, a third person, the operator, would just come into the conversation without being invited, and this would always be funny. The operator was listening in to the conversation. Just like the old telephone system, a reflex point belongs to a special wiring where everything is directly connected.

[66] The conversion occurs in an area called the reticular formation of the brain.

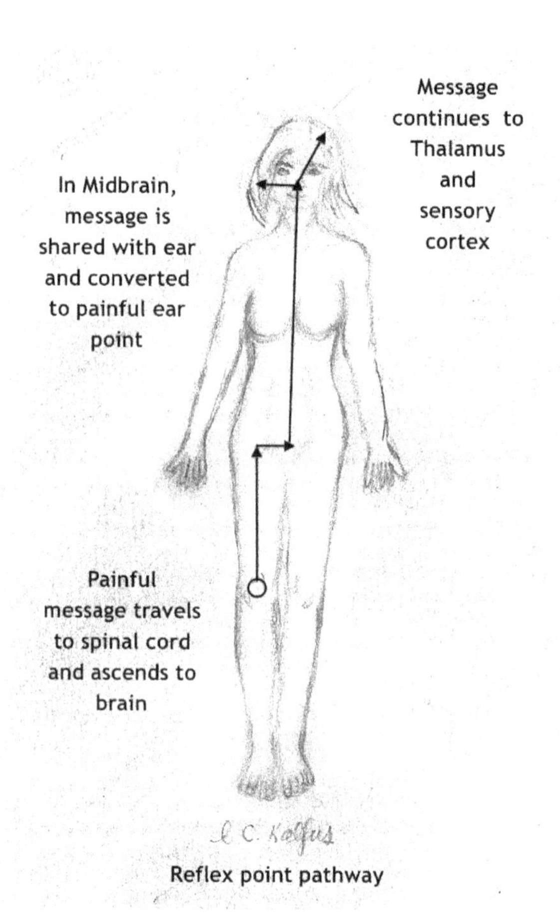


Example of Reflex point pathway: painful message begins at knee and travels to spinal cord and up to the brain. Before message reaches upper levels of brain for interpretation, message from knee is shared with the ear and converted to painful ear point.

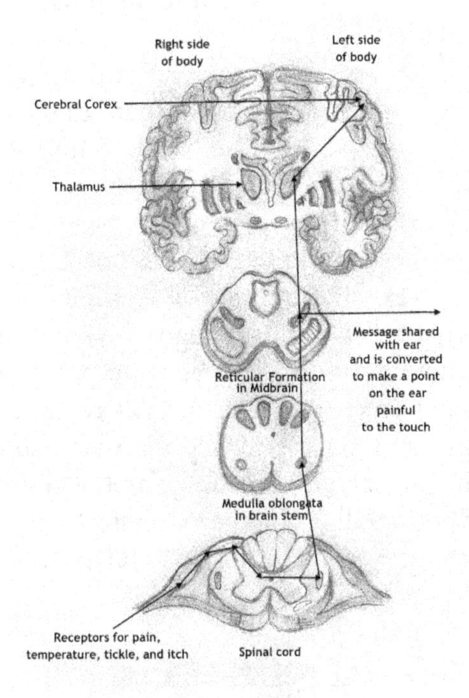


Reflex point pathway in spinal cord and brain

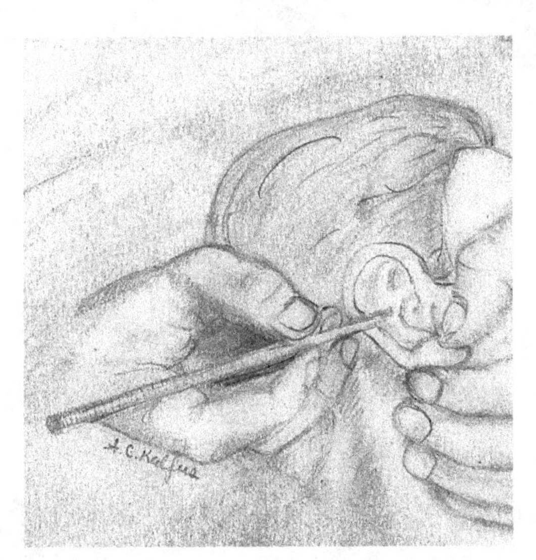

Pain message that originated in the body is converted to painful ear point

## A simple way to elicit ear pain

Dr. Paul Nogier developed an elegant, reproducible and indisputable method to demonstrate how pain in the body provokes pain in a corresponding area of the ear due to the way our nervous system is wired. To begin the experiment, the ear should be found to have no painful response to the gentle pressure with a probe in the area where the experiment is to be performed, even with repeated attempts. Next, the thumb, for example, of the person is pinched. Unlike just a few moments ago, the same area of the ear that corresponds to the thumb will become progressively painful while the areas adjacent to the thumb area remain silent. As the duration of the pressure on the thumb increases, so does the pain on the corresponding ear point. Finally, as the thumb pressure is removed, the ear point will become progressively less sensitive until it is completely insensitive to a probe, even after repeated attempts. See the illustrations below.

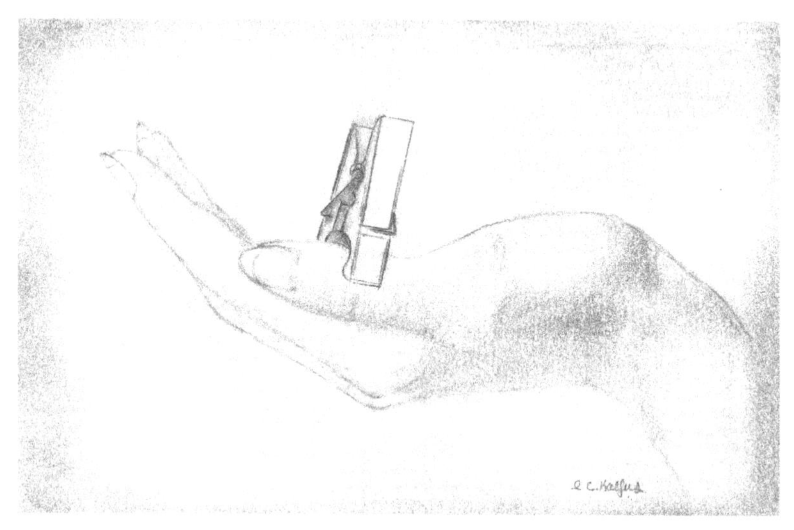

Evoked pain on thumb

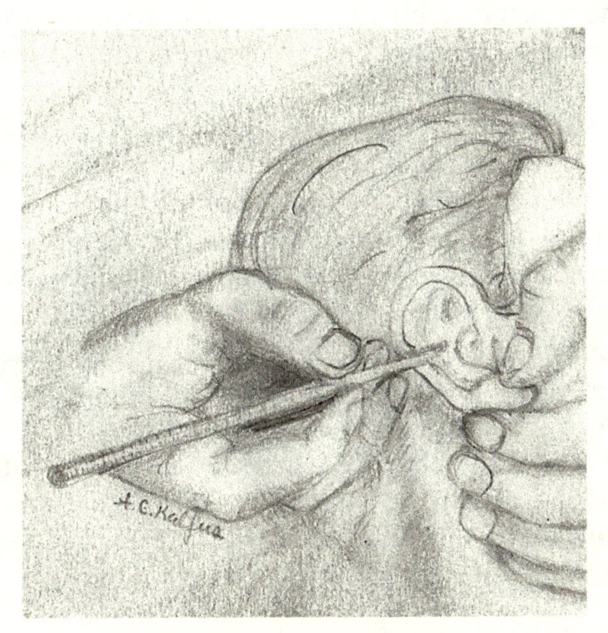

Examining for painful point on ear

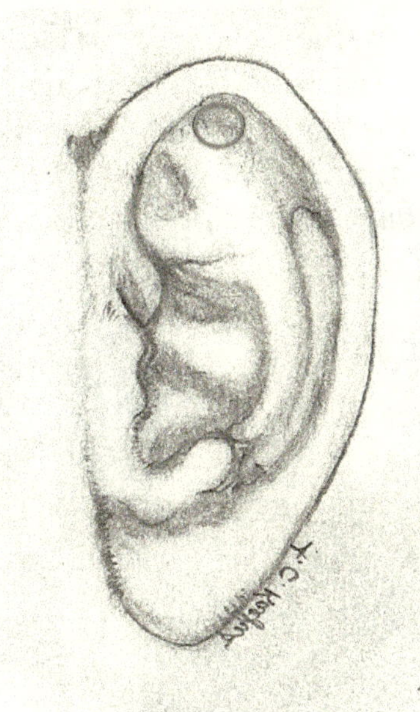

Painful "reflex point" on ear corresponding to evoked pain on thumb

This demonstration clearly shows that *a painful area of the body can induce a corresponding painful area on the ear.* The level of pain experienced in the body is also proportional to the level of pain experienced on the corresponding ear point. The sensitivity (or disappearance of sensitivity) to touch on this point is proportional to the duration (or removal) of the stimulus (thumb clamp). This is one example, of many, that shows a quality that is *unique only to the ear; that a point or area will become "active" only when there is a corresponding body pain or pathology. Otherwise, when healthy, the ear will remain inactive or insensitive.* Note that 80 percent of the time, the ear point that corresponds to a painful area on the body is on the *same side* that the pain is coming from. In other words, pain from the left thumb will likely result in a painful ear point on the corresponding left ear. Twenty percent of the time, however, the opposite ear (in this example, the right ear) will demonstrate a corresponding painful ear point.

## Electrical detection of an ear point

As mentioned above, the second way to locate an ear point is by electrical detection. Unlike the rest of our body, our ear will produce electricity *only* when there is an underlying pathology. And the *only scientifically proven way to detect electrically active points on the ear is by the use of a device that detects differences in the flow of electricity.*[67]

[67] See diagram below of a bipolar probe.

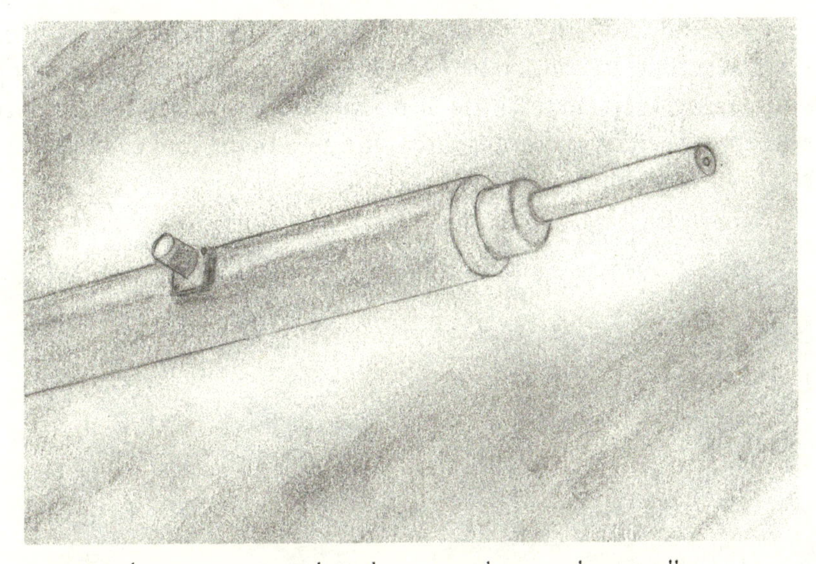

Instrument used to detect and treat electrically active ear points

This means that the flow of electricity at a point on the ear must be different from the flow of electricity on the area immediately surrounding that point. The device will show precisely where there is a drop in resistance to the flow of electricity signaling an underlying problem in a corresponding area in the body. A bipolar electrode places *two* electrodes against the skin of the ear and compares how conductive or resistant they are to the flow of electricity. The points that have the greatest differential in the flow of electricity are the most valuable points to treat. Bipolar electrical detection instruments measure three areas of electrical resistance:

1. The skin's resistance *to the flow of electricity* at a small point in the center of the wand/probe.
2. The skin's resistance to the flow of electricity in the area surrounding the small point in the center of the wand.
3. The skin's resistance to the flow of electricity on the patient's hand, which is holding an electrical ground.

If there is little or no difference in the flow of electricity between these points, there is no value in treating the point. In the illustration below, the small area of the inner stylus would be touching the skin of the ear and will have either a lot of electrical activity or very little electrical activity in relation to the larger outer area that's surrounding the inner stylus. If there's little or no difference in electrical activity between the center and outer areas, the area being looked at doesn't need to be treated.

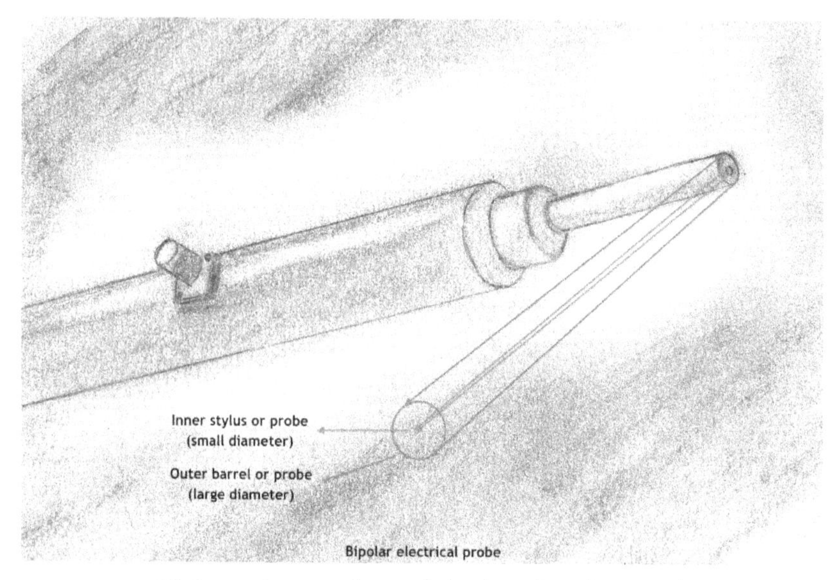


Enlarged view of tip of the bipolar electrical detection and treatment instrument

There are some factors that my colleagues and I have observed that make electrical detection more difficult, especially for very complex exams. One is use of cortisone (over 10 mg). Also, *detection seems to be most accurate in the morning rather than in the afternoon.*[68] Additionally, when atmospheric pressure is very low (stormy weather), points of the skin tend to become very difficult to differentiate by electrical detection. So when I do electrical detection and I find several points I say, "ah-ha, a storm is coming up," which is meant to be funny.

---

68 Due to something we call chronobiology, which is our body's natural cycle or rhythm (i.e., sleep cycle) that's affected by the rhythms of the sun and moon.

## Four distinct characteristics of each active ear point

Electrically active ear points have very unique characteristics. Researchers have discovered that *all ear points that have a low resistance to the flow of electricity have four distinct features: an artery, a small vein, a nerve, and a lymphatic duct.* All of these structures are very close together, represent something active, and have all the properties to emit electricity. It's because of this electrical emission that we are under the impression that the resistance to the flow of electricity at the level of the skin has dropped. Thus, *the points on the ear that are detected to have a low resistance to the flow of electricity are very functional complexes and when they are active, they emit electricity.* The detection devices used to locate these points simply determine if a point allows a tiny electrical current to pass through it (conductive) or if it is resistant to the flow of electricity emitted from the device. But *these points actually produce electricity,* so we are under the impression that they are less resistant to the flow of electricity. When we find a point that has low electrical resistance, we know that underneath it is an active complex of blood vessels, nerve endings, and an immune system duct producing electricity, and it only does this when there is an underlying pathology in the area of our body that the point corresponds to.

## Our ear helps to regulate our body temperature

In addition to emitting electricity, electrically active ear points also play a role in the *regulation of your body temperature.* This is called *thermoregulation.* An area of the brain known as *the hypothalamus regulates temperature control* (and thirst), and it's located very high up in the brain. A good way to think about this is to use the example of a high-rise building. At the top, we have the boss's office. She likes the temperature set to be very cool. Imagine that the entire buildings temperature is controlled by the boss's thermostat on the top floor. I can assure you that the temperature will not be the same on every floor because the thermoregulation on the top floor will not be the same as the bottom floor. Your body is about the same as the building. At the top, you have the thermostat in the brain. And if you want to regulate the temperature at every level of the body, you will need a thermostat at every level. And so in the ear, the electrically active ear points (i.e., neurovascular complex points) are the points that will help regulate your body temperature at every floor. The ear points simply start regulating the temperatures of the areas of the body that the points correspond to when the balanced temperature is offset. For example, if you place one leg in an ice bucket, the ear point that corresponds to your leg will start working. Likewise, if you place your arm into something hot, the point that corresponds to you arm will also start working and become electrically active. So what does an increase in temperature mean? It means that an area in your body is becoming more active and doing more work. And work is related to function. When we are assessing a point for being electrically active, we are actually assessing the function of the corresponding organ. We are not able to determine if the organ is painful, rather whether the organ is functioning or not functioning normally. Thus, for example, if we detect the stomach point on the ear, we can conclude that the function of the stomach is not normal. This is because *electrical activity is actually measuring temperature control, which is measuring the function of a particular organ.* These points are very sensitive, and thus very responsive to little particles of light known as photons. The light photons can be red, infrared, or include colors such as blue, green, and yellow. The ear points can also be treated with very specific *pulses of light known as the Nogier frequencies.* If neither photons of specific colors of light nor pulsating frequencies of light are available, tiny needles can treat the points.

The particles of light used in auricular therapy are very soft. Our ear reacts to very small stimulation. Just simple light; white or colored light. I use frequency-dependent red and infrared light, which is a fantastic "weapon" against functional disorders.[69] Light is also very effective against postural disorders,[70] and immune disorders. We have very good results in helping immune disorders with auricular therapy when using light or electrical frequencies to treat active ear points, but not with needles.

---

[69] Areas of our body that are not working well.

[70] Leading to dizziness, headache, chest, kidney, stomach and lower back pain, sinus problems, and chronic fatigue.

## The ear and its immune system guardians

Earlier, I explained that the points on the ear that are electrically active (emitting electricity) have four unique structures: An artery, a vein, a nerve, and a lymphatic. But where does this lymphatic duct go? It connects to the closest lymphatic node. As a reminder, our lymphatic system helps us to get rid of toxins and waste and other unwanted materials. And is it random that there's a lymphatic duct on every ear point? Is it just there to enhance and embellish the points? No. Their presence means that *each point is hooked up to the lymphatic system and the immune system.* A special type of immune cell called the *Langerhans cell* surrounds every single ear point and acupuncture point. These cells are like guardians and play a very important role in our immune system by stimulating other cells to protect against infections in the skin. Because the immune system is hooked up to each ear point, we have very good effects in helping immune disorders with auricular therapy.

When we treat an electrically active ear point with a needle, we destroy the small artery, vein, nerve, and lymphatic associated with that point. Remarkably, in sixty-two hours (or three days), this complex rebuilds itself and is raised up again, and it rises up just next to the point that was punctured. This means that when a treatment is repeated with needles, it will not be effective in the exact same point because it has moved slightly. We will need to wait for at least three days before we can treat the same area again for it to be effective.

## Treating the ear using electrical stimulation

Although electrical treatment can be used on an electrically active ear point, it isn't necessary because these points are very sensitive and responsive to very low energy light. Using electrical therapy on the ear points has been likened to driving a tractor on the highway; it can be driven there, but it's not meant to be there.

In summary, ear points that have a low resistance to the flow of electricity (electrically active) are best treated with gentle pulses of light, which I refer to as frequencies (i.e., photons of light). Points that are sensitive or painful to the touch and are not electrically active are best treated with tiny needles. For those points that are both sensitive to touch and electrically active, both treatment methods can be used.

# Chapter 13

## In the Beginning
## (The Origins of the Frequencies Found in All the Animal Kingdom)

In elementary school, we were studying the origins of the universe. We learned that some scientists believe that everything began with one enormous explosion of energy and light. This is called the Big Bang Theory. I also learned from the book of Genesis that said everything began when GOD created the heavens and the earth.

And GOD said, "... Let there be light." Although I was being taught two completely different points of view, they did share one thing in common: That in the beginning, there was light.

That same day when I first learned about the beginning of time, I went home and asked my family, "What do you believe happened in the beginning? Did everything in the universe just start with a 'bang' or did GOD create everything?" My parents quietly sat me down and said that all of life is a miraculous gift and this gift is continuous in all whom the spark of life is kindled. And they left it at that; they never did answer my question. Instead, I was encouraged to learn as much as possible about both points of view and in time, they said, the answer would find me. So here I am, years later, writing *Wave Catcher*, which has a lot to do with the therapeutic benefits of light, the same light that was created and given to all of us in the beginning, and thanks to the discoveries by Drs. Paul and Raphael Nogier and the supportive research conducted by several colleagues, I am able to help relieve pain and suffering simply by using this miraculous gift.

## The light of life

If there is a single chapter in this book that best explains the concept of light therapy and the reasons why our body is so responsive to light, this is it. At first, I thought about using the phrase "Once upon a time ..." as the introduction since what I'm about to share with you is a narrative that few people outside of a select group of scientists have ever heard. However, I didn't want you to think about this as a literary fairytale. Instead, I want to share with you a story that took many dedicated doctors and researchers their lifetime to piece together. And as recently as April of 2016, scientists from Northwestern University in Chicago have captured images of a flash of light that sparks at the moment of human conception. These flashes of light are now seen in humans for the first time and can be used to determine the viability of an individual egg or embryo for couples relying on in vitro-fertilization treatments. Clearly, my mom was ahead of her time when she shared with me the historical teachings that said all of life is a miraculous gift and this gift is continuous in all whom the spark of life is kindled.

This chapter will explain to you why our skin, and in particular the skin that covers our ears, is capable of adapting to our needs by capturing only the frequencies of light it needs to stimulate, regulate, and repair itself.

## The seven frequencies of light that connect all animals on earth

There are seven frequencies of light that every animal on earth responds too. Human beings, however, are the only animals that respond to an eighth frequency. What this means is that when different areas of our body and/or ear are exposed to specific frequencies of pulsed light, there is a momentary change in the strength of our pulse. But why does our body respond to light this way? And how can these different frequencies be used to help us? This chapter will answer both of these questions.

Our story begins when the earth was very young. And what's remarkable is that today, the *same* frequencies of light that bathed the earth and all living things at the beginning of time, can be used to *speed up our natural healing process.* It's this *light of life* that forms the basis of auricular medicine.

## The story of cells and their love of light

In the beginning, when some forms of life started as a single cell, only minerals, water, and light bathed the earth. At this time, *sunlight* was the only form of energy, and a single cell could detect this light and position itself toward a particular frequency that today is known as the Nogier frequency "A." *Frequency A is defined as a light pulsing on and off at a rate of 2.28 times per second.* A single cell can actually "see" the light with a very small set of tubes called Centrioles, which can map the direction of the light source. It can also weigh and process vast numbers of signals from outside and inside its body to make complex decisions. Thus, a single cell is by no means a simple cell. In time, as this cell joins together with other cells to become a more complex organism, we find that these groupings become responsive to a different frequency of light. Nonetheless, we must always keep in mind that *no matter how complex an organism becomes, its development all began with frequency "A."* As a result, *any single cell or group of individual cells that is damaged or stressed will readily recognize and respond to this frequency. Its therapeutic value can be seen in any type of cellular damage such as in inflammation, swelling, burns, or infection.*

As cells gather together and are no longer "single," they have to develop some features that weren't necessary as single cells, such as sharing nutrition and information. Now, each cell needs to recognize its neighbors and work together. *Frequency "B"* fosters all the chemical reactions that support the *physical interaction between the cells* because multiple cells need to *communicate and coordinate their actions together.* It *is defined as a light pulsing on and off at a rate of 4.56 times per second. "B" frequency supports the immune system, metabolism, and the absorption of nutrients.* This frequency is often referred to as the woman's frequency because many of the most common diseases in women are nutritional (absorption of nutrients) and immune disorders.[71]

In order to carry nutrients from one area to another, we need to develop a system of *muscle movements.* These movements are referred to as *motility.* An analogy of this is seen when we squeeze a tube of toothpaste and the paste is moved through the tube while the tube itself doesn't actually move. *Frequency "C" supports all the cells involved in the movement of nutrients or waste to include the cells of the heart, lungs, arteries, and digestive system. "C" frequency is defined as a light that is pulsing on and off at the rate of 9.12 times per second.* Conditions that affect muscular movement[72] will be treated with this frequency. Frequency "C" also stimulates the release of the hormone dopamine[73] that helps control the brains pleasure and reward center.

The fourth step in development of an organism is its ability to travel. *For an organism to move from one place to another, it will need two things: muscles and the development of symmetry.* In every animal that moves, there is symmetry and its body can be divided into two parts that are similar. *If symmetry doesn't exist, there is no possibility*

---

[71] That is 80 percent of autoimmune disorders are found in women.

[72] That is muscular tremors such as those seen with Parkinson's disease.

[73] The same hormone that is lacking in those with Parkinson's disease.

*of movement.* An organism must be able to separate between its left and its right. In the previous stage of development, muscle cells were formed to provide for the movement and the exchange of nutrients or waste. Now, *in order to move, an organism must have an exact evenness or uniformity when it is divided into halves.*[74] The organism must learn to move its body parts around a central axis or reference point in a coordinated manner to be able to move effectively. The nervous system is organized to do just this. Every organism that can move from one point to another by itself has a nervous system that is divided into two similar halves. The brain is divided into two hemispheres, each side controlling one half of the body. The two hemispheres are connected and communicate with each other by a broad band of nerves. This is a very important relationship that has been overlooked in the past. *If symmetry doesn't exist, there is no possibility of movement. This symmetry develops using frequency "D" from the sunlight, which acts on the nerve fibers that interconnect the two cerebral (brain) hemispheres. "D" frequency is defined as a light pulsing on and off at the rate of 18.25 times per second. Conditions that affect or limit your movements*[75] *can be treated with "D" frequency.*

As an organism continues to develop and begins to travel, it needs a simple nervous system to help control its movements and react quickly to environmental stresses. *Frequency "E"* is now *captured from the sunlight* and *develops the Reflex System with the spinal cord. It is therapeutic for reducing pain and stimulating the spinal column. "E" frequency is defined as a light pulsing on and off at the rate of 36.50 times per second.*

*Frequency "F" fosters the development of the lower brain,* which is the area below the cortex (i.e., the subcortex). Although known as the *fundamental frequency for scar healing,* it is also used in a broad range of treatments to include the *healing of wounds, ulcers, and bone fractures.* Furthermore, it has *therapeutic effects on emotional disorders (antidepressant), appetite regulation, and on hormones.*[76] Frequency "F" is often referred to as a women's frequency[77] due to its

74 A honeycomb, a sunflower, a snowflake are examples of uniformity.

75 That is uneven walking or poor posture.

76 Hypothalamus of the brain.

77 Along with frequency "B."

affects on nutrient absorption, feelings and sensations, and emotions. *"F" frequency is defined as a light pulsing on and off at the rate of 73.0 times per second.*

The highest level of development in an organism is fostered by *frequency* "G." This is the frequency that *stimulates* the cortex, or *the higher brain* system. It manages more complex functions such as *intellectual thought, complex voluntary movements, emotions, psychosomatic issues, memories, and language.* It also has therapeutic value in the *treatment of chronic pain. "G" Frequency is defined as a light pulsing on and off at the rate of 146 times per second.*

At the end of this chapter, there are illustrations that show where on our body and on our ear each of these frequencies is "captured" by our skin.

## Our skin is like that of a Chameleon

Dr. Paul Nogier discovered that when a particular area of the body is exposed to a particular frequency of light, there is a momentary and unconscious change in the strength of the pulse (heartbeat). For thirty years, he studied how his patients' pulse reacted when different areas of their skin was exposed to white light, colored light, infrared, and electrical frequencies. Dr. Nogier discovered that it is the *frequency* or *rate at which a light is pulsing on and off that is most significant rather than the color (or lack of color) of the light source.* In a healthy person, our pulse will respond only to a single frequency of light on a particular area of our body and will not respond[78] to

---

[78] No change in pulse strength.

other frequencies. When that same area of our body is injured or ill, our skin in that area (and the corresponding area on the ear) will now become responsive (change in pulse strength) to different frequencies of light until that area is restored to health. Thus, *our skin is like that of a chameleon.* However, instead of changing its color or patterns, *our skin will change what frequencies it captures to adapt to the needs of our body.* Frequency therapy can then be provided to accelerate the rate of repair. *Light* frequency happens to be *the most powerful way to treat an electrically conductive point in the ear* and the healing language of light is also the gentlest, most familiar, and most readily recognized therapy used in auricular medicine. Furthermore, it does not bring with it a list of adverse side effects.

The chameleon-like nature of our skin provides us with sensitivity to all sorts of frequencies. Our body can be likened to an electrical power bar that is hooked up to the sun and the stars. As mentioned above, our skin will change what frequencies it captures to adapt to the needs of our body. "A" frequency stimulates our cells; "B" frequency stimulates our respiratory and digestive systems and supports the immune system, metabolism, and the absorption of nutrients. "C" frequency stimulates muscular movements to share nutrients; "D" frequency stimulates our system of symmetry and movement for locomotion; "E" stimulates our complex muscle movements and reflexes; "F" stimulates our lower brain and affects feelings, sensations, and emotions; "G" stimulates the upper brain for our most complex brain functions, which includes chronic pain. How does all of this help us to feel better if we are not healthy? Here is an example to show you how these light frequencies actually work. Let's say you injure your shoulder during an accident. Normally, your shoulder captures frequency "C," but now, we see that the skin around the injured shoulder captures frequency "A" as well. The reason why "A" is now causing a momentary change in the pulse is because the *skin is designed to capture light that stimulates, regulates, and accelerates the system of repair in our body.* Your ear, in particular, will capture light frequencies that are particular to your pathology, and we can use your ear to accelerate this system of repair. In this example, we can treat the injured shoulder simply by exposing the corresponding ear point to frequency "A."

*Light is the most powerful system to treat ear points and only the least electrically resistant points will receive this light therapy.*[79] Light therapy at these specific points in the ear will provide results over time that will last much longer and provide a much deeper affect than with needles. Treatment needs only to be three to four seconds, rather than minutes.

## In the complexity of a problem, is the simplicity of the solution

Light therapy is actually very simple. Since we know what frequencies different areas of the body and ear normally respond to,[80] we can readily locate a problem if a particular area is responding to additional frequencies. For example, when a person is healthy, any point in the area of the middle area ear will react only to the B frequency.

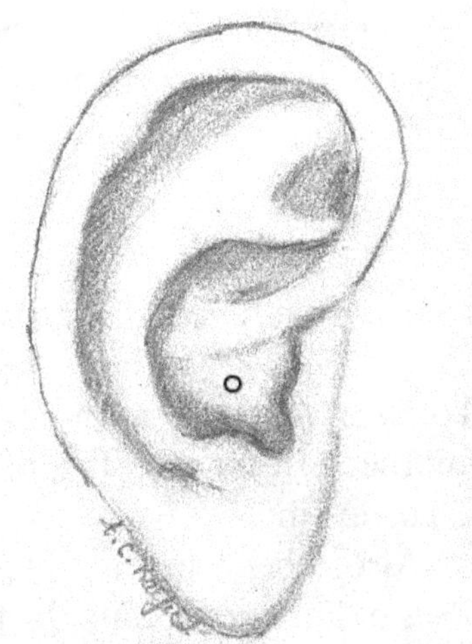

This point primarily reacts to the B Frequency

[79] Not the points that are painful to the touch.

[80] As measured by an increase in pulse strength.

However, if any point in this middle area also reacts to additional frequencies (such as A, C, and E), the point should simply be treated with these frequencies. Upon treatment with these additional frequencies, the point and the area in the body that the point corresponds to, will return to a healthy state and will only respond to its normal "base" frequency.

In summary, different areas of our body respond to only certain frequencies of pulsed light. Sometimes it's easier to think of the different frequencies in terms of the areas of our body that they relate to. Area "A" is located near the opening of our ear and matches to all of the openings of the body.[81] All these area's respond[82] to the "A" frequency of light. Area "B" is the front part of chest and abdomen area. Area "C" is the upper and lower limbs (arms and legs). Area "D" is a stripe-like area that starts at the pubic area and goes up and along the midline, over our head and back down our back to the our bottom. "E" area is around the neck and then follows an area around the vertebrae until our first lumbar vertebrae. "F" area is the face area and a stripe-like area on the back of the scalp. "G" area is the forehead to the top of the scalp and the sides of the nostrils. The "L" area, discovered by Dr. Raphael Nogier, is an area *unique to human beings* while the other seven frequencies can be found in all of the animal kingdom. It is the frequency for *laterality* disorders[83] that include learning or behavioral problems, dyslexia, stuttering, and other conditions.

## The frequencies are related mathematically

Interestingly, all of the frequencies found in the animal kingdom are related to one another in that each is a multiple of two of the previous frequency. For example, to find frequency "B," multiply frequency "A" by two. Also, when we multiply frequency "G," which is 146 on and off pulses of light per second, by two, we get the same harmonic frequency effect as we had with "A." Thus, both the 2.28

---

[81] That is ears, eyes, mouth, nose, urinary.

[82] As measured by a momentary increase in pulse strength.

[83] Imbalances between the left and right hemispheres of the brain.

frequency and a 292 frequency (146 × 2 = 292) can be used for "A" frequency with the same effect. The only exception to this rule is *frequency "L"* which is an *extra frequency only found in human beings.*

The diagrams below illustrate where on our body each frequency will, under healthy conditions, elicit a subtle increase in our pulse strength. The illustrations will also show where, on our ear, the same frequency will elicit the same response. Under healthy conditions, only one frequency will cause a change in our pulse strength while all the other frequencies will elicit no response and have no effect on our pulse. When we are experiencing a problem in any area of our body, additional frequencies will elicit changes in our pulse strength. It's these responses to additional frequencies that I refer to as the unique "signature" of a particular problem.

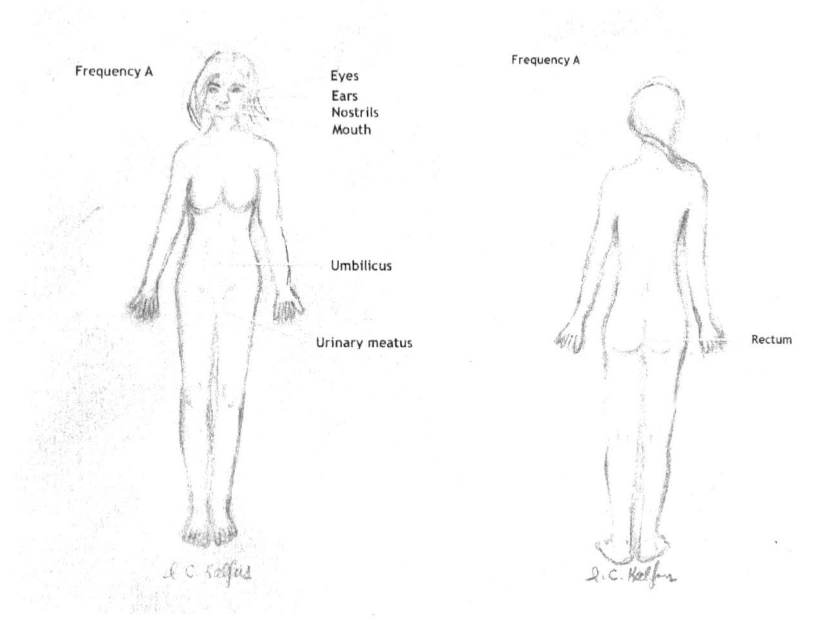


All body openings respond to frequency "A" @ 2.28 Hz

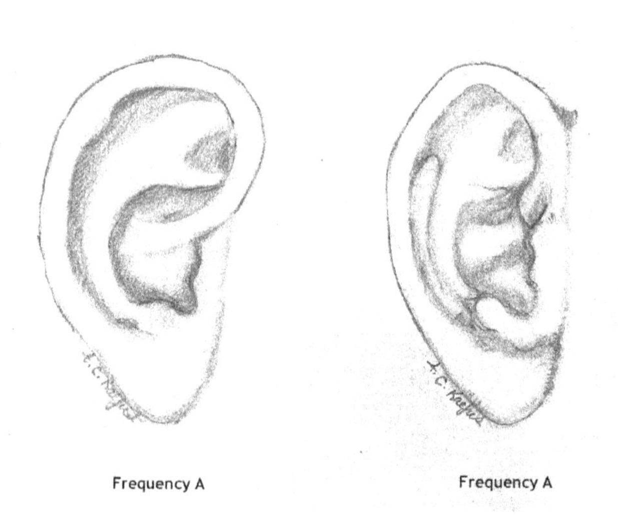


Corresponding area on the ear, near ear opening, that responds to frequency "A"

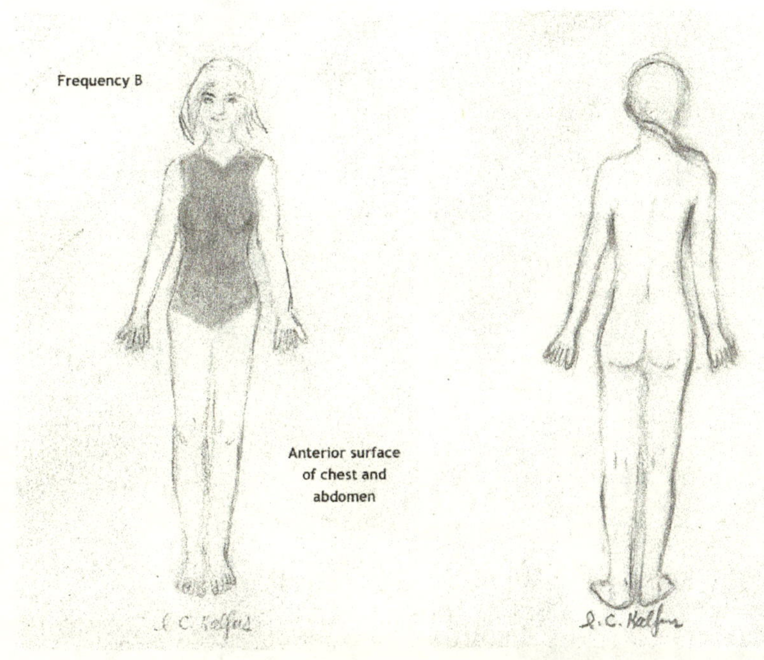


Front areas of the chest and abdomen respond
to frequency "B" @ 4.56 Hz

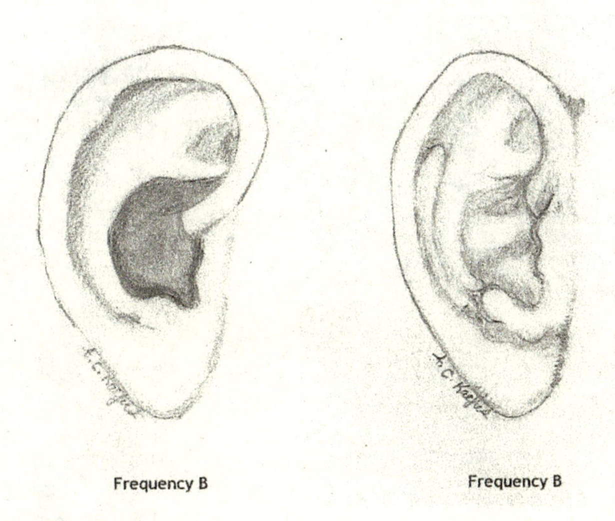


Corresponding area on the ear that responds to frequency "B"

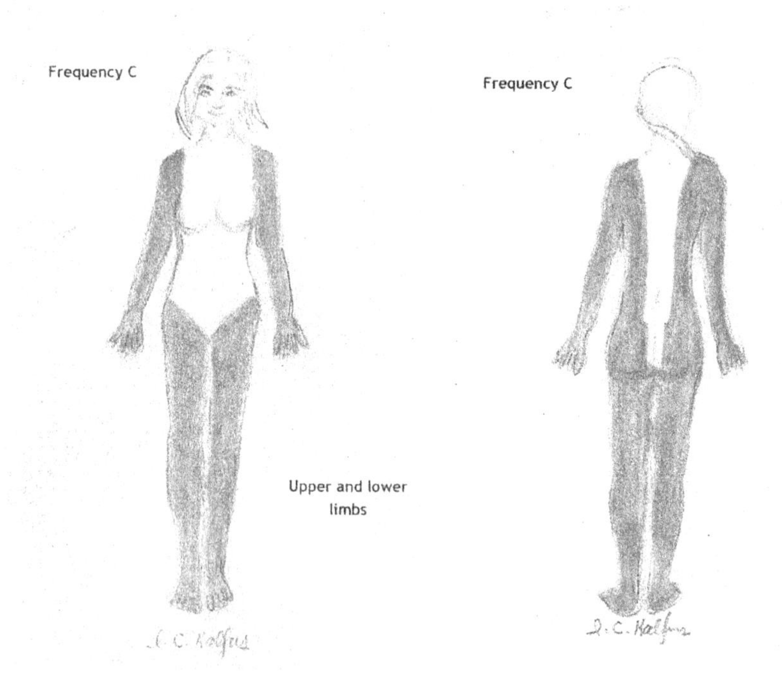


Upper and lower limbs respond to frequency "C" @ 9.125 Hz

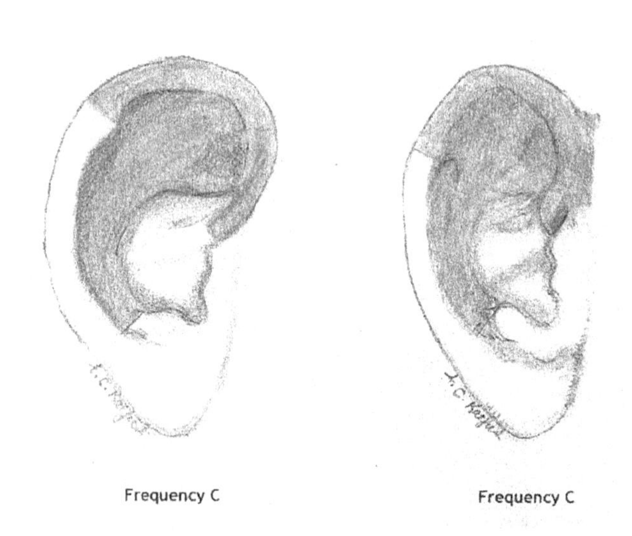


Corresponding areas of ear, to include the rim, that respond to frequency "C"

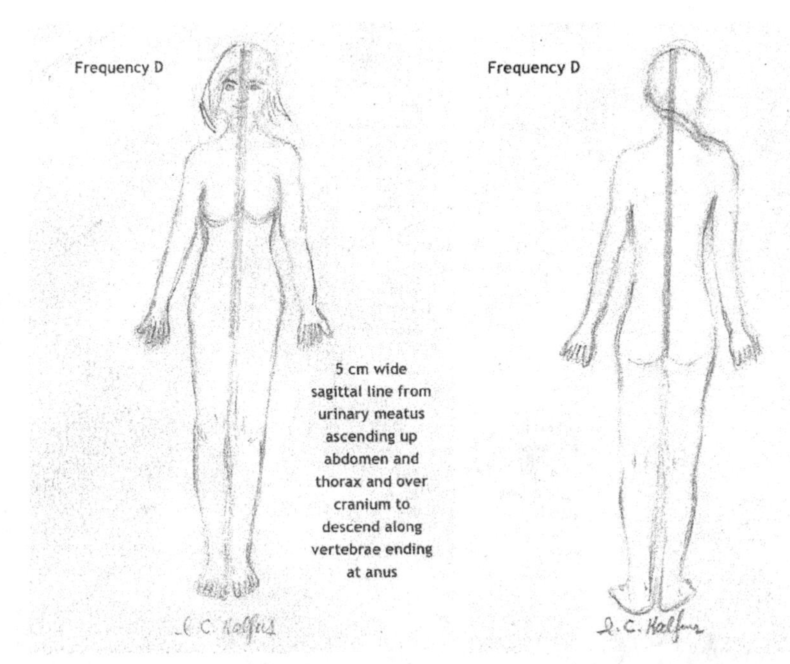


Midline area in front and back responds to frequency "D" @ 18.25 Hz

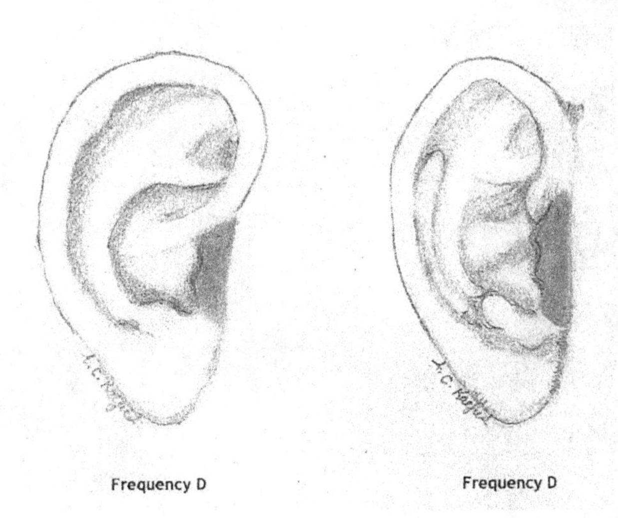


Corresponding area of the ear that responds to frequency "D"

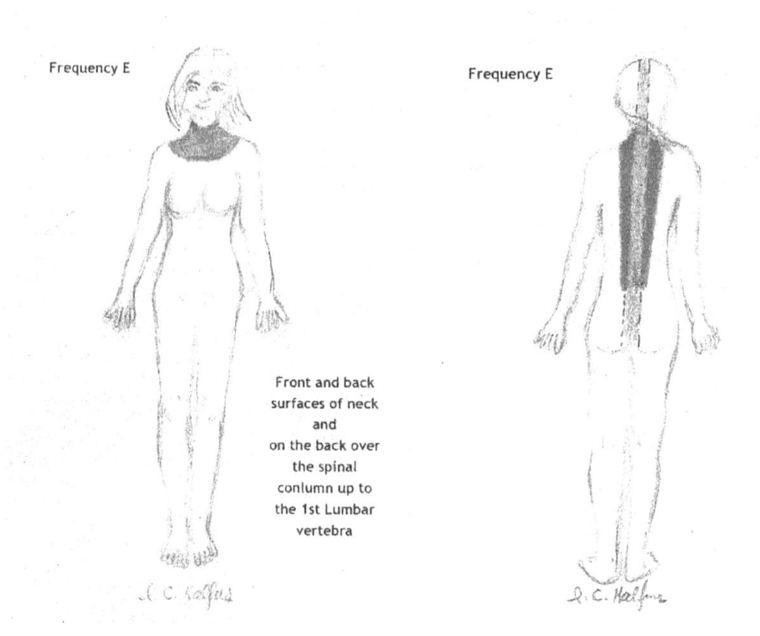


Front and back of neck, and over spinal column respond to Frequency "E"

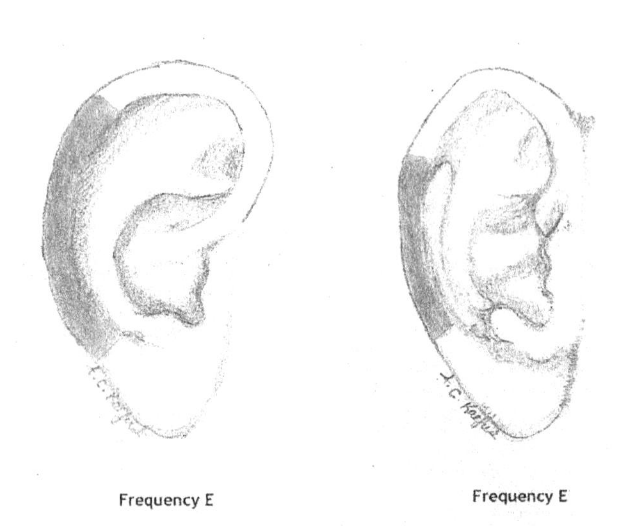


Corresponding area of the ear that responds to frequency "E" @ 36.5 Hz

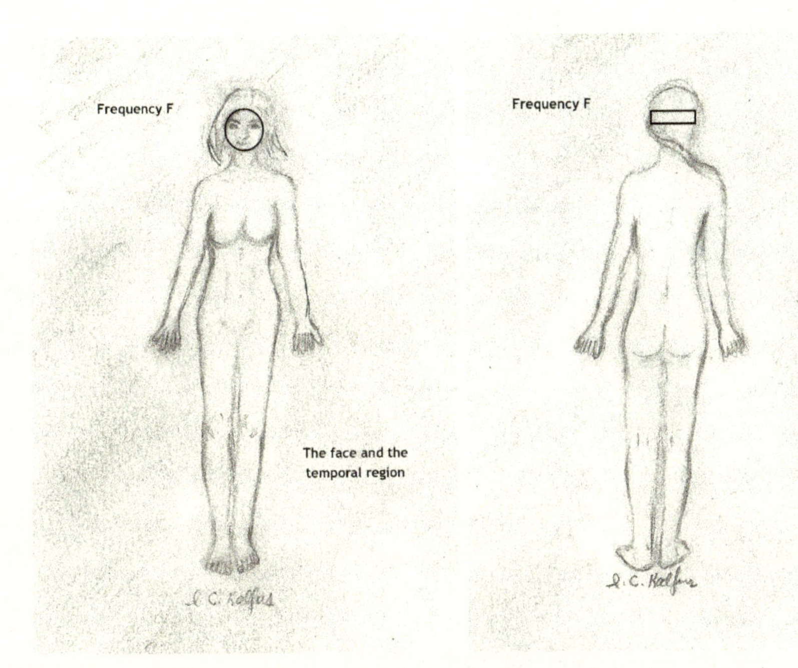


The face and an area along the back of the lower scalp respond to frequency "F'

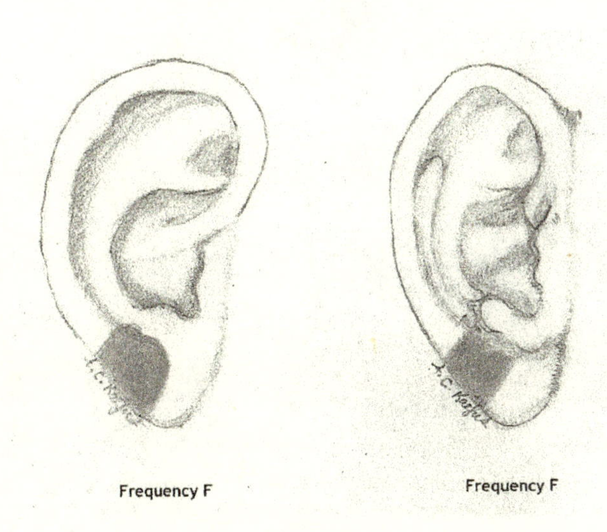


Corresponding area of the earlobe that responds to frequency "F" @ 73 Hz

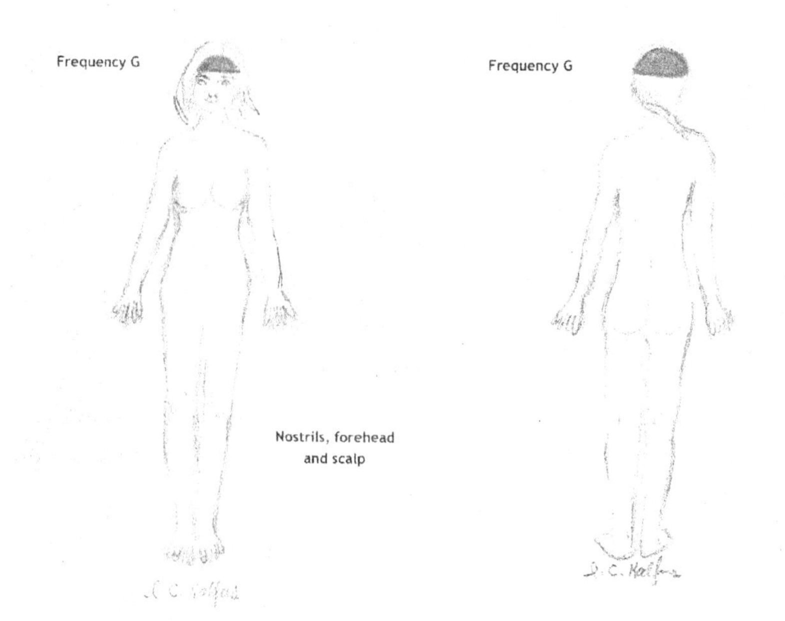


Forehead, back of upper scalp, and rim of the nostrils respond to frequency "G"

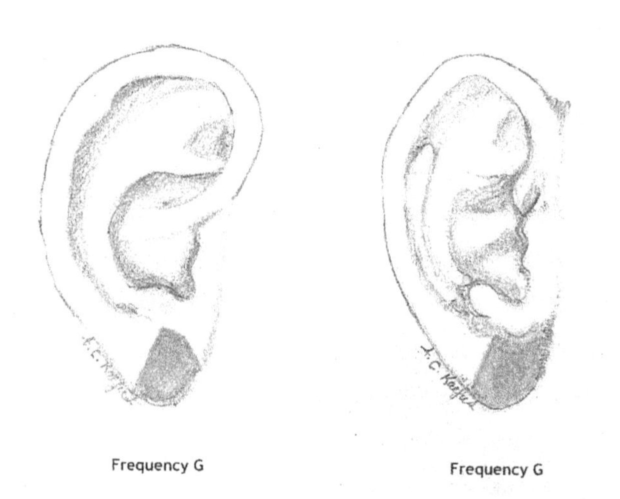


Corresponding area of ear lobe that responds to frequency "G" @ 146 Hz

## A frequency that sets people apart

One of the most difficult groups of disorders to treat successfully with auricular therapy are those that occur when there is an imbalance between the left side of the brain and the right side of the brain. Under healthy conditions, the two sides cooperate together. Under unhealthy conditions, the two sides compete against one another, which can lead to learning disabilities, attention deficit, and depression, to name a few. Although we're able to improve a person's posture or balance for walking by improving the communication between to two sides of the brain, we weren't having much success with laterality disorders. Laterality disorders occur when the left and right halves of the brain are competing for control of the body rather than working together. This causes interference of one side of the brain by the other side.

Laterality disorders are most often found in left-handed or ambidextrous people, and they occur in 10–20 percent of the population according to Nogier. *No other species apart from human beings are either right or left-handed.* Animals do not display a dominant side.[84] Even chimpanzees, which most resemble us, use either their right or left hands and feet with no apparent difficulty. On the other hand, even when newborn, human beings show a tendency toward one side or the other with ninety-eight percent of adults being right-handed and possessing a dominant left half of the brain.[85]

Raphael Nogier continued to study frequencies of the body and the ear, which led to the discovery of an additional frequency that's unique to human beings. He named this frequency "L" for laterality and it resulted in predictable therapeutic success. The area of the body that responds to this frequency is located on the midline of the skull from where the hairline begins in the front near the forehead and continues to where the hairline ends in the back of the skull near the neck. See the illustrations below.

---

84 No hemispherical lateralization.

85 Coutte and Zorn 1999.

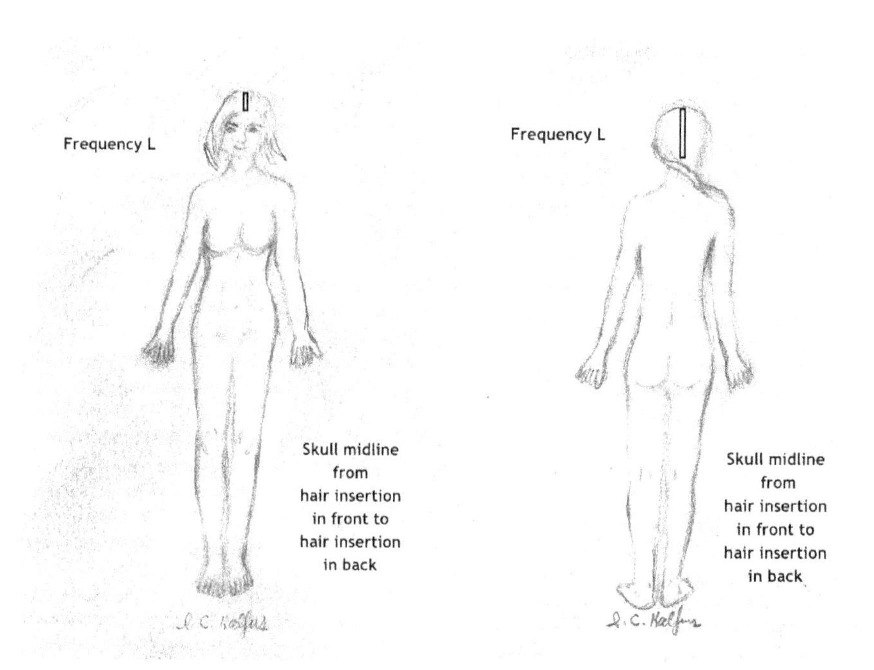


Midline of skull area in front and back
responds to frequency "L" @ 276 Hz

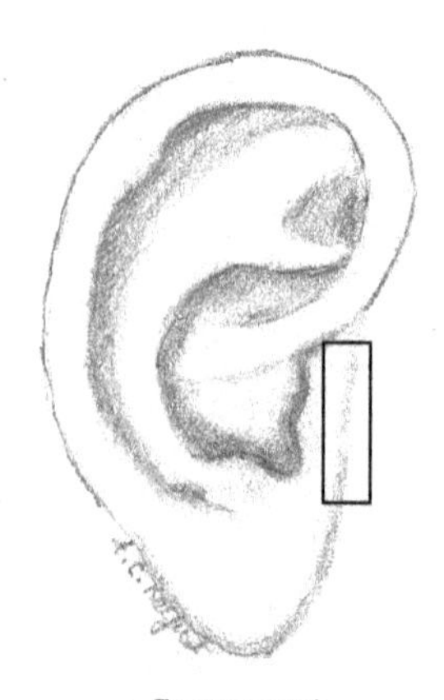

Frequency L

Corresponding area on ear near cheek
responds to frequency "L'

When this area is exposed to the "L" frequency, which is defined as a light pulsing on and off at the rate of 276 times per second, an increase in the strength of the pulse is detected under healthy conditions.

Scientists have long known that hyperactivity of the right side of the brain can result in hypersensitivity, the inability to stay still, anxiety, and depression. Likewise, we have known that hyperactivity of the left side of the brain can result in a loss of practical reality.[86] It's possible to detect blocks in laterality[87] by stimulating specific areas of both the left and right ear with specific frequencies while simultaneously noting any changes in the strength of the pulse. Note that *the right ear corresponds to the left side of the brain while the left ear corresponds to the right side of the brain.* Typically, the dominant ear, which has more electrically active points, will correspond to the dominant side of the body (i.e., the right ear for a right handed person).

## Precise diagnosis of left and right brain communication disorders

This topic is one of the most difficult to understand in the field of auricular therapy. In the past, we haven't had much success in treating disorders that arose when both sides of the brain were in competition with each other. Today, however, there's much more clarity on this subject.

Raphael Nogier discovered that when there is balance between the two halves of the brain, a light frequency of 8.74[88] along the outer rim of the right ear[89] will cause an increase in pulse strength, while a frequency of 3.79[90] on the right ear will *not* cause an increase in pulse strength. Also, the frequency of 3.79[91] along the outer rim of left ear

86 No longer in touch with reality.

87 Blocks in the communication balance between the left and right sides of the brain.

88 Known as the high frequency for laterality.

89 For a right-side dominant person.

90 Known as the low frequency for laterality.

91 Idib.

will cause an increase in pulse strength, while a frequency of 8.76 will not. This shows that, for a right-handed person, the right side is "balanced" by the left side. In a true left-handed person, the above is reversed. A good way to think about this is to compare the frequencies of light to weights on a weighing scale. If both sides are equal, the scale will remain balanced. If either side is unequal, the scale will tip. The two diagnostic frequencies that Dr. Nogier discovered allow for a more precise diagnosis as to the type of laterality disorder.[92] *Treatment for laterality disorders will be found on the ear and the precise points will be located by electrical detection.*

## Humanity has a dominant side

According to Dr. Nogier, ninety percent of people are truly right handed with their language centers in their left half of their brain while only one percent are truly left handed with their language center in the right hemisphere. The remaining nine percent are poorly lateralized lefties with their language center in the left hemisphere. Humanity, therefore, has a dominant side, which we refer to as *laterality.* If the left half of the brain is dominant, you will be right handed and if the right half of the brain is dominant, you will be left handed. If for example, you are right handed and were unable to use your right hand, a laterality disturbance may occur (i.e., if right handed, throwing a ball with your left hand will feel and look strange).

A "bridge" of over two hundred million nerves attaches the two halves of the brain. This bridge is called the *corpus callosum* and a problem here is usually the cause of a laterality dysfunction. *Communication between the halves is critical in determining how a person functions.* If communication is broken, the two halves have to work on their own which makes it difficult for a person to have an all around perspective. Such confusion or mixed dominance between the two hemispheres may result from intrauterine or birth stress, injury during development, and physical, emotional, or psychological

---

92 That is neurological diseases such as multiple sclerosis, depression, and attention deficit disorder.

stress. When this happens, there will be *considerable confusion in the individual.* This is because one half of our brain is dominant and logical while the other half is nondominant and emotional. *The dominant side prevails in verbal and language functions, while the nondominant side is important in understanding the nonverbal cues or "body language" and intonations of a person's voice. There needs to be a balance between the logical person and the emotional person to have an all around perspective.* Laterality points on the ear can only be detected electrically and only when there is a dysfunction. *Treating any active laterality points on the ear prior to other treatments will help to improve the overall results of auricular therapy.* Otherwise, the *laterality issues will present a significant blockage to healing.*

We need a balance between our logical person
and our emotional person

The symptoms of laterality most often appear by the age of seven and include dyslexia, learning difficulties, memory disorders, poor concentration, stuttering, and attention deficit disorders. As adolescents and adults, they may also include nervousness, hyperactivity, gastrointestinal dysfunctions, problems with coordination, and depression.

To determine the laterality or dominant side of a person, the most common way is to see which hand they use to write with. Another way is the hand-clapping test. Here, the right-handed person will likely use his left hand more passively. The direction of a person's gaze when they're thinking is also an indicator of a person's dominant side. See diagram below.

The direction of gaze is deviated to the side opposite the dominant hemisphere

## Less is better

It's important for the doctor to treat any electrically active points on the outer rim of the ear first and then move toward the center of the ear. Any point treated will change the entire dynamic of the ear. A peripheral (outside) point will "turn off" points that are more central in the ear while a central point will "turn on" points peripheral to it leading to multiple false and non-therapeutic treatment points. *We want to treat as few points as possible with the least amount of frequency.* Too much information (treatment) destroys information. *In auricular therapy, less is better.*

## Light and depression

A recent study by Schiffer et al., 2009, demonstrated the effects that light has on the treatment of patients diagnosed with major depression and anxiety, which are both laterality disorders due to hyperactivity of the right hemisphere. This study suggests that light therapy on the forehead alone may be beneficial for the treatment of laterality disorders. Light therapy is also routinely used at the Mayo Clinic, which is the largest medical group practice in the world, to treat seasonal affective disorders (seasonal depression) as well as a host of additional conditions. For more than twenty years, clinicians have used bright light therapy to successfully treat seasonal affective disorder. Light therapy, which uses a full spectrum of frequencies to simulate sunlight minus the ultraviolent light, has far fewer side effects than medications and can provide relief within days. In a meta-analysis (which combines the results of multiple studies) completed in 2005, the American Psychiatric Association indicated that bright light therapy alone was an effective treatment for both seasonal and nonseasonal depression, and was as effective as medications.[93]

Light therapy for the treatment of depression

93 Golden RN, Gaynes BN, Ekstrom RD, et al. The efficacy of light therapy in the treatment of mood disorders: a review and meta-analysis of the evidence. The American Journal of Psychiatry. Apr 2005;162(4):656-662.

In summary, the origins of the frequencies found in all the animal kingdom can be traced back to our beginnings. We followed the journey of an irreducibly complex single cell (i.e., the "simplest" form of life) and watched it develop into an even more complex multicellular organism. Throughout this process, the organism became enriched by the sun and selectively captured different frequencies of light at different stages of development. Today, we are bathed in the same *light*, which is like a *language* that every living thing on earth understands and responds to. *Our skin, and in particular the skin that covers our ears, is capable of adapting to our needs by capturing only the frequencies it needs to stimulate, regulate, and repair itself. Once received, these frequencies will speed up the healing process.*

# Chapter 14

## Obstacles in the Treatment of Chronic Pain

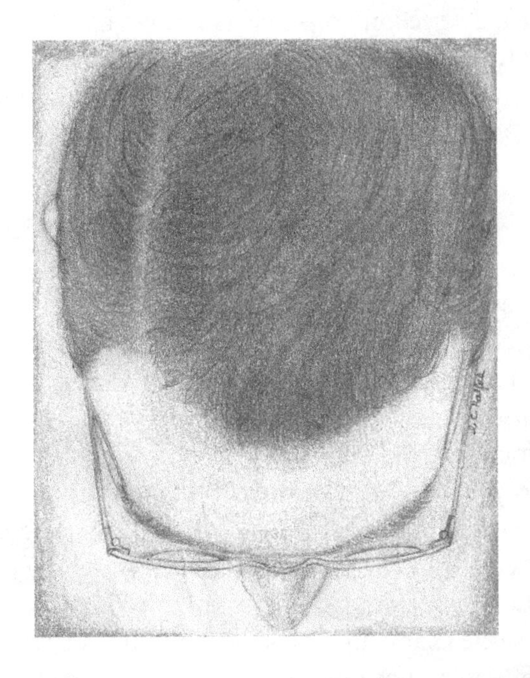

I've talked about how your body receives light from the environment and then interprets the information that's coded within the light to help stimulate, regulate, and speed up your own natural healing process. In fact, I've all but said that without sunlight, life, as we know it, wouldn't exist. But what happens when you receive information, and there are obstacles that block it from being delivered? It can be both frustrating and disappointing when treatment that should work, doesn't. And this is what happens when doctors begin treatment without first ruling out and then removing any barriers that exist. This also misleads some practitioners to think that auricular therapy is not predictable and it's a hit or miss treatment.

This chapter will discuss the most common obstacles to auricular therapy as well as how to discover them, and then treat them. Once removed, treatment will be predicable, successful and rewarding.

There are two important obstacles in the treatment of chronic pain that are commonly overlooked. Each of these is like a torpedo that causes widespread destruction to therapies, has dramatic consequences, and often results in a disappointing outcome. The first obstacle is known as the *First Rib Syndrome* or the Stellate Ganglion syndrome, which revolves around a group of nerves that travel along the spinal cord. *The* second obstacle is known as a *Dental Focus* and it revolves around a tooth or dental-related pathology. By addressing them, your ailment may be resolved quickly. *By ignoring them, a problem may persist indefinitely regardless of therapy.*

## First rib syndrome

First rib syndrome, which is also called Stellate Ganglion syndrome, affects 15 percent of the population and *90 percent of the time, it affects the person's left side.*[94] This ganglion, or *collection of nerves, is located on the back of your neck/shoulder area* and is on each side of your spinal cord. Your first rib connects in two places: in the front onto your sternum or breastbone and in the back between the seventh cervical and first thoracic vertebrae. The enlarged end of the rib, called the head, connects into the vertebrae just below the stellate ganglion on each side of the spinal cord. The ganglion are not the exact same shape on each side. They are very small (1/4"–¾"), and they belong to a chain of nerves that run along the spine called the sympathetic chain. This chain uses the hormone *adrenaline* and it's possible *after experiencing trauma*[95] *that the head of the first rib*

94 The remaining 10 percent that affect the person's right side are typically left handed.

95 Such as lifting a heavy weight, sitting, or sleeping in a position for a long time, auto-accident.

*would press on the ganglion. The first rib syndrome always starts from a particular event.* This collection of nerves at the Stellate ganglion affects several areas in your body including your heart, lungs and airways, arms, eyes, and your brain. When there is an *uplifting of the first rib* (usually the result of a specific event), there will be an *adrenaline release* in all these areas, and it will create a disorder that is very violent and long term; the *first rib syndrome is adrenaline-driven.*

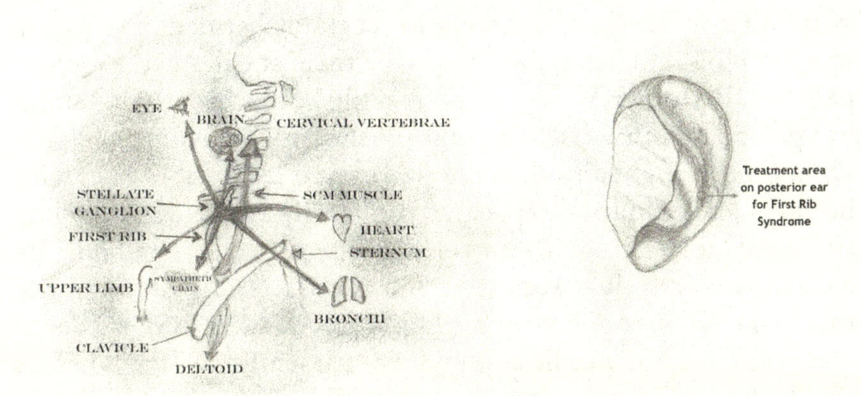


The first rib pushes up into a collection of nerves that refers pain to distant areas including the shoulder, arm, airways, heart, brain. and eyes

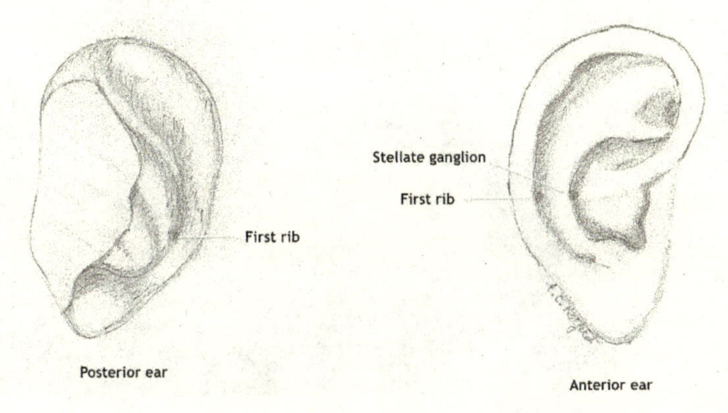


Treatment areas for First Rib/Stellate Ganglion Syndrome

To better understand just how disruptive and wide spread the effects of First Rib syndrome can be, I will share a few actual patient treatments and outcomes with you.

The first story is about a colleague of mine, a physician, who came to see me with a shoulder pain that was radiating up her neck and giving her daily headaches. After years of consultations, medication, and physical therapy, her problem still remained. I asked her if there was a particular moment when she first felt the shoulder pain. She said, "Yes." About ten years ago, she was swimming in her pool and went to reach for her daughter as she was jumping into her arms. Ever since then, she has had shoulder pain that has progressed to give her daily headaches. Upon taking her pulse on both wrists simultaneously, they were normal when she looked straight ahead. When she moved her head to her right, her right pulse almost disappeared completely and came back when she looked forward again. She is left-handed. To treat her, I quickly located the area that was electrically active (and painful to the touch) on the back of her right ear and placed a tiny ASP needle in that point. An ASP needle is a small French semi-permanent needle that looks similar to an earring and it falls out on its own in three to four days.

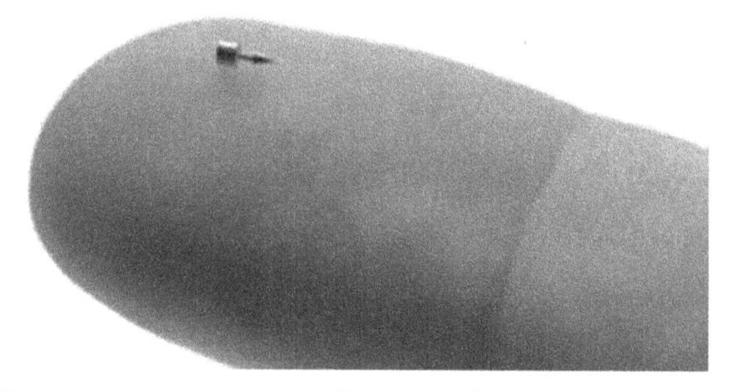

French semi-permanent needle on my fingertip

I always look to treat the back of the ear first for muscular relief with First Rib syndrome because I want to drop the first rib by relaxing the muscles. I also placed a needle on her other ear that was electrically conductive. I checked her pulse, and there was still a slight drop even though her pain was gone. I checked again and found a sensitive point to the touch on the front part of her right ear and placed another tiny ASP needle. All her pain was now gone and her pulse no longer disappeared when she turned her head to either side. Ten years of shoulder pain and associated headaches were gone in a few minutes. I saw her on a follow-up visit six weeks later and there was still no pain—First Rib Syndrome.

The second story is slightly more complicated and not what you would expect from First Rib syndrome. This patient came to me for a consultation to see if something in her mouth/her teeth was causing a reaction making her feel nauseous and sending her to the bathroom several times a day (almost every hour). She went to see many specialists. All tests were normal. She had previously ruled out food allergies and after a few appointments with me, it was clear that she wasn't improving and I couldn't find anything. I said that I was sorry, and that I think we she should stop at this point because I couldn't find anything wrong with her. She was very kind, and I wanted to help her, but I couldn't continue without finding a problem to resolve. As she was leaving, she made a gesture with her arm and shoulder, and I asked her why she was doing that particular motion and she said, "Because it hurt there." I asked her to go back into the room so I could take a look. It was the First Rib syndrome on the right side. *This is rare because they are almost always found on the patient's left side.* She called me that next day to tell me her problem was completely resolved. She returned for a recall a few months later and she had no problems. Everything returned to normal.

The third story is even more complex with an even more refined diagnosis. A patient came to me after a history of slipping and falling down. She had been seen by multiple doctors, and they were unable to resolve her problem. Usually after questioning a patient, we can gather a lot of information. She said that normally when she feels like fainting, she lays down to prevent herself from falling. She remembered a specific time and day when she first noticed the dizziness. She was on a long trip with friends on a bus. She slept on the bus, and the next morning when she woke up, she felt bad and actually fell onto the floor of the bus. An ambulance came, drove her to the hospital, but they didn't find anything wrong with her. I asked her how she slept. She said she slept on the seat, she didn't sleep well, and her head was leaning to her left side. When she awoke, her head was still leaning to her left and when she went to stand up, she fell to the floor. This was First Rib syndrome. Within minutes, I detected and treated a point on the back of her left ear (to relieve the muscular tension) that was both painful to the touch and electrically active. I also located a point on the front of her left ear that required treatment. She no longer slips or falls down and her pain is gone. The treatment was simple and it was her ear that guided me to the needs of her body.

As you can see, when the head of the first rib is displaced upwards following physical stress or a traumatic event, the displacement of the rib mechanically irritates and compresses this collection of nerves called the stellate ganglion. This gives rise to a wide range of problems such as instability, ringing of the ears, headaches, visual problems, blood pressure problems, feelings of chest compression, feeling of upper limbs getting weak, trigeminal neuralgia, frequent urgent intestinal problems, and cardiac problems (changes in heart rhythm).

## How to diagnose a First Rib syndrome

To reach the diagnosis of a First Rib syndrome is simple. First, there's a precise moment or date when the problems first began and all of the treatments and tests that we've received up to now have failed to help. Even if a person arrives complaining of a toothache but also has a first rib syndrome, the pain will continue to persist following dental treatment. As mentioned earlier, *ninety percent of the time, the First Rib syndrome will be on the patient's left side and on the occasion that it is on the patient's right side, the person is usually left handed. When there is a compression in the left stellate ganglion due to the first rib pushing into it, the pulse of the left side will be lessened and feel more distant.* What I will do is *take the pulse of my patient bilaterally (both hands at the same time) and if the pulse is less on one side than the other, there is stellate ganglion compression on that side.* Sometimes the first rib syndrome appears normal when the patient is looking straight ahead but changes when she turns her head to the left, to the right or toward the sky. I look for *dissymmetry (unevenness) between the two pulses.* I take the pulse with my thumbs on both wrists simultaneously. The reason I use my thumbs is because the tip of the thumb contains the greatest number and the highest density of nerve receptors of all the fingers. So don't believe any bad reputation the thumb might have as a pulse detector. This is a technique developed in France by Dr. Paul Nogier, and further refined by his son Dr. Raphael Nogier who personally mentored me.

A second characteristic of a first rib syndrome can be observed from the patient's back. I can see an asymmetry when I ask the patient to breathe deeply. I look for a difference in mobility between the left and the right first rib joints. One joint will be more mobile than the other. There is also pain upon palpation of the first rib joint that is less mobile and slightly out of place. It's important to know that *there is no need to physically manipulate the rib back into position because auricular therapy can easily do it with only one or two ear points. There is a single point on the back of the ear that is the most important point and it can readily be located by either electrical detection or by finding the point that is most painful to the touch. The back of the ear will affect the muscles that attach to the first rib and will return it to its normal position.* This will immediately relieve the pressure that the first rib was placing on the stellate ganglion. Once treated, I will again take my patient's pulse on both wrists at the same time and find that there's no longer a difference between the two; they are both the same. Usually, this simple therapy resolves most of the symptoms and "viola," you feel good again.

## Dental focus

A second important obstacle to the treatment of chronic pain that must also be addressed and ruled out is known as a *Dental Focus*; and it revolves around a tooth or dental-related pathology. Interestingly, *50 percent of all shoulder pain* and 90 percent of Achilles heel problems are associated with a dental focus, according to Dr. Nogier's findings. And if a dental focus is not addressed and treated, we will not be able to correctly treat shoulder or Achilles heel pain.

*Whenever a person comes to me with a chronic shoulder or heel pain, I will always look for a dental focus.*

So what is a dental focus? It's defined a *detrimental message* that originates from a tooth or areas in the mouth related to our teeth, such as the sinus area or the structures that support our teeth. The cause of this *damaging message* may be visible (an infection or a broken tooth) or nonvisible, painful or nonpainful, visible on images (i.e., X-rays) or not, a current or past problem, and which elicits at a distance another pain or disease that apparently has no relationship to our teeth. A dental focus can elicit several signs and symptoms such as fatigue, functional disorders of all kinds,[96] urinary tract infections, and generalized pain (like shoulder pain). Usually, *pain associated with a dental focus is repetitive and chronic.*[97] *The dental focus is one of the most difficult problems to diagnose in auricular therapy.* Since half of all shoulder pain is associated with a dental focus, and the best way to find a dental focus is for the practitioner to properly "read" the pulse while examining the roots of the teeth, it's critical that the practitioner is skilled and able to recognize the subtle changes in the pulse on the wrist.[98] Thus, *before treating any shoulder pain, we must first look for a dental focus.* Regardless of the area in the mouth that the damaging message is coming from, *the outer ear provides us with access points to neutralize the message.* And these *treatment points are located on the lower border of your ear lobe* as seen in the illustrations below.

---

96 Different areas of our body that don't work properly.

97 Headaches, shoulder pain, and lower back pain.

98 To detect a vascular autonomic signal.

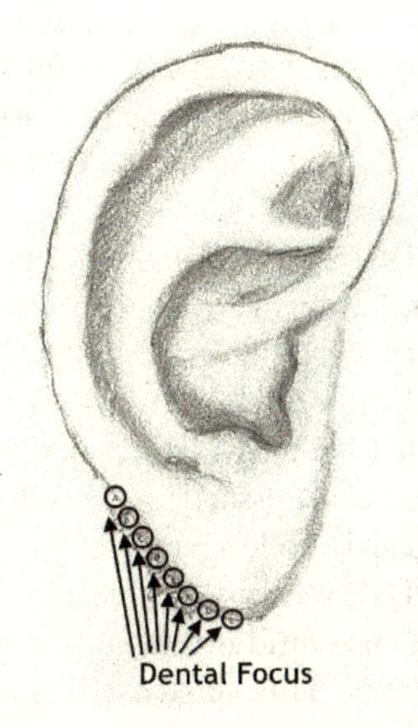


A dental Focus is an inflammatory message best treated along the peripheral ear lobe.

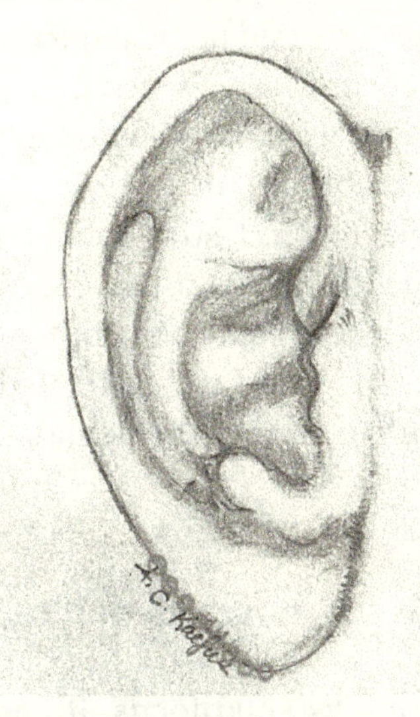

Dental focus treatment points are located on lower border of the ear lobe and are found by electrical detection.

## A painful message can remain long after its source is gone

Imagine a tooth with a problem, such as a wisdom tooth or a tooth that was treated and is damaged. Our tooth may be physically or chemically damaged and sends out information that reaches one or two targets in our body. A dental focus is a message that emanates information to one of several targets. And even when the source of the message is no longer there,[99] the *information remains* and can be serious and detrimental in terms of chronic pain. For example, if I were to ask an audience to stand up while I was speaking to them, they would likely stand. Now, imagine that I am withdrawn from the room, taken away while they were standing. They will likely remain standing. This is because the message to stand still remains, although the source of the message is no longer there.

Here is a story of a twenty-three-year-old woman who was referred to me because she could not sleep through the night due to vomiting for several years. This started when she was very young. There was no cause or disease found for why this was happening. She was sent to me, and I immediately found a dental focus on a *left tooth* and *treated an electrically active point on her right ear along the ridge of her lower ear lobe.* I treated her just once and on the following day, she said she did not vomit. A few months later, she returned for a follow-up and everything was fine; she no longer vomited at night and sleeps without interruption. To confirm that the source of her vomiting was indeed her tooth, I stimulated the root area of that particular tooth with a gentle light through the cheek area. She called me the following morning and said she vomited for the first time in several months. I saw her the next day, and treated her and everything was fine again. This confirmed that the dental focus was indeed the source of her problem.

## How to diagnose a dental focus

Here is how I look for a dental focus. It's actually very simple. While taking the pulse of a patient, I gently push on the roots of each

---

[99] The damaged tooth is restored or removed.

tooth through the cheek and lip areas. Normally, the pulse remains stable. If the strength of the pulse changes, I know that there's a dental focus. It's as simple as that. When I feel the pulse strength change, I count how many stronger pulses there are. I usually detect two to three signals (a small focus) but some practitioners have detected as many as nine. *If the focus were on the left side, I would treat the right ear on the edge of the ear lobe. If the focus were on the right side, I would treat the left ear. The only way to find the focus point on the ear lobe is with electrical detection. If a practitioner is unable to take the pulse and detect any changes in its strength, an electrical detector on the opposite ear lobe will help locate the point.* And since there is a corresponding area of pain in your body (i.e., shoulder pain), I can place a tiny needle in the point or treat with frequency and the shoulder pain, for example, is likely treated. If any pain remains, then I will simply locate the point in the ear that is painful to the touch and that corresponds to the shoulder point. Once located, I can simply place another tiny needle in that point or treat it with frequency (either light or electrical stimulation). It's that simple. Just a few precise treatment points on the ear are all that is needed. If light detection is available, it can be used to detect a dental focus thru the cheek instead of pushing on the roots of each tooth. In a healthy state, tissue will respond normally to light and there will not any change in pulse strength. However, in an unhealthy state, damaged tissue (dental focus) will not respond normally to light and there will be an increase in pulse strength when in the presence of specific light frequencies. Through the cheek and lip area, a gentle light is directed to the root area of each tooth. I will look at my device that emits light and see at what frequency the tooth is responding, as measured by changes in the pulse strength. And remember that a change indicates an unhealthy state or dental focus. If it is the tooth that's responsible for the shoulder pain, for example, it will respond to the same light frequencies in the same way as the painful shoulder. Once the frequencies of light that cause a change in pulse strength[100] are noted, I will look for the exact same pulse response to the same frequencies on the lower border of the opposite side ear lobe. Then, for further precision, I will locate the point on the ear lobe

100 When directed on the root of a tooth.

with electrical detection and treat with either a tiny needle if there is pain associated with the focus, or with frequency[101] if there is no pain. Remember that *for acute pain, tiny needles can be used but for chronic pain, light therapy is best.* To verify that the proper point was treated, I again check the roots of the teeth by either pushing on them or using light while simultaneously checking the pulse for changes. If the pulse remains the same, perfection.

*Generally, by treating the dental focus, the shoulder pain will disappear.* However, if shoulder pain remains, there will likely be a sensitive point to the touch at the shoulder point in the ear on the same side as the pain (right shoulder pain, right ear point). If so, it's easily located and treated with a tiny needle. We can also directly scan the area of the body that hurts (i.e., shoulder or foot) with frequencies of light and evaluate if a particular frequency causes a change in pulse strength. Each pathology or painful area will have its own unique "signature" of light sensitive frequencies. Once determined, we can see if a tooth also shares the exact same light signature.[102] If so, we have precisely located a dental focus. For example, if your pulse strength changes when your right shoulder is exposed to light frequencies A, B, and C, and a dental focus also causes the same changes in pulse strength when exposed to the same light frequencies A, B, and C, then I will match this light signature or pattern to an area on the opposite side ear lobe that also responds to A, B, and C. Once a matching point is found of the opposite ear lobe, I will treat this area knowing that the dental focus was the precise cause of the shoulder pain. *By treating a dental focus* with either tiny needles or frequencies, *we are stopping the harmful message from being broadcasted throughout your body.* If there is problem in the tooth or sinus area, this too must be resolved.

In summary, before I begin to treat pain patients, especially those with shoulder or Achilles heel pain, I will rule out major blockages to treatment, which include the first rib syndrome and a dental focus. Once these are ruled out and treated as needed, the remaining treatments will likely succeed.

---

[101] Light or electrical stimulation.

[102] Responses to different light frequencies.

## The Nogier Pulse

One of the keys to auricular medicine is the ability of the doctor to take and interpret the pulse. When Dr. Nogier first taught me his method, he impressed upon me just how fine and delicate the signal is. In general, it takes several months of practice before a person can begin to be proficient at it. And without proficiency, a practitioner could misinterpret the vascular signals.

The Nogier pulse is found on the wrist and is a reflection of involuntary reactions that our arteries undergo when our skin is stimulated by either touch, pressure, white light, pulsed lights, colored, or infrared lights. These reactions can be detected by feeling a change in the strength of the pulse. It feels similar to a tire that is being pumped up with air. To demonstrate just how delicate this signal can be, let me compare it to an experience that most of us are familiar with. Imagine that you are at a party with several guests. Everyone around is talking, and the room you're in is filled with noise. In order for the person next to you to hear what you're saying, you have to raise your voice a bit. Suddenly, the hostess of the party interrupts you and says, "I hear the baby crying upstairs." You could only hear all the noise in the room but the attentive mother, unconsciously paying attention to her child in the other room upstairs, was able to hear and recognize her crying baby despite all the noise. In this analogy, the noise in the room corresponds to the normal pulse, the one that we as healthcare providers learned in school. The Nogier pulse is the cry of the baby. To distinguish between the two is only a matter of touch and a lot of practice. So it's important for the practitioner to be patient and attentive.

Practitioners thumb on patient's wrist to monitor pulse

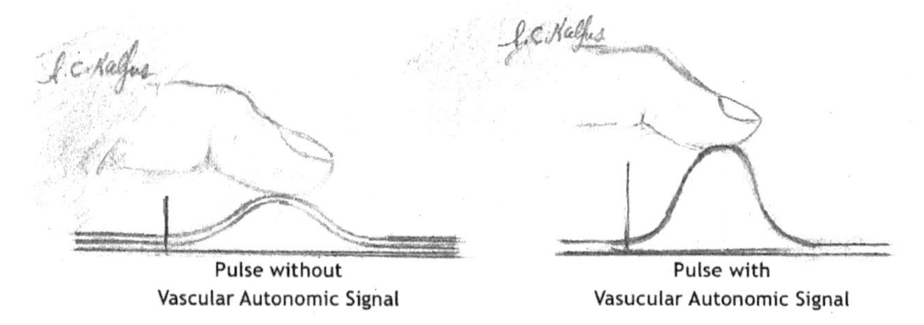


Visual representation of the Nogier pulse

# Chapter 15

## Look for What Is and Not for What Should Be

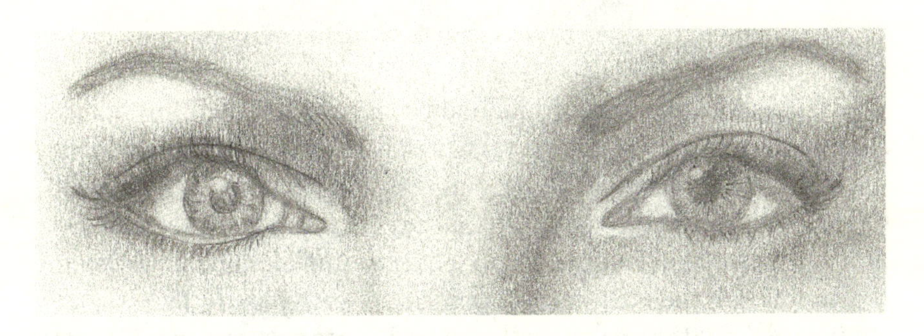

Before Paul Nogier first presented his observations, he discovered an unexpected connection between the outer ear and the entire body. This led to his first great insight of "the man in the ear" which today is referred to as a *Microsystem.* Microsystems are like small images of the entire body, located in different areas on our body. Our ear is a complete microsystem that's unique in that it has the same *orderly anatomic arrangement as our body* and has been *shown through research to be the most effective microsystem.*

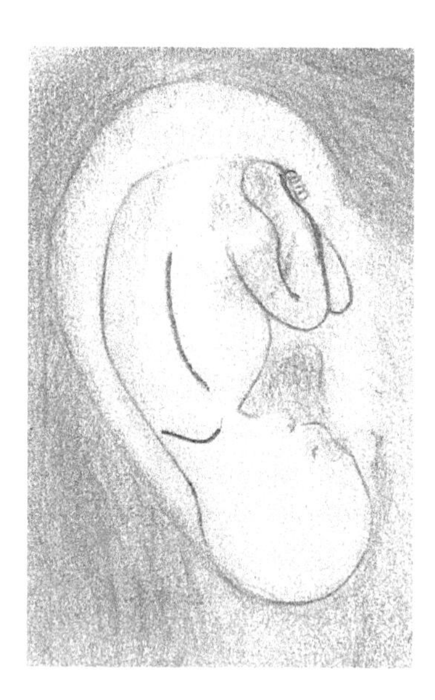

Unlike any other area on your body, your ear points will only become electrically active (emit electricity) when there's an underlying problem. Thus, your ear serves as a guide to your health. The World Health Organization announced in 1990 that auricular therapy was, "the most developed and best documented, scientifically, of all the micro-systems of acupuncture and is the most practical and widely used." This is due to the type of tissue that makes up the ear and due to its direct connection to the nervous system, which controls every single cell and bodily function.

Your ear not only has the most complex surface anatomy and shape with all of its folds, ridges, and valleys, but it also has the highest concentration of electrically active points[103] comprising over ten percent of all body acupuncture points. Additionally, your ear is the only place on your entire body where each of the three basic developmental cell layers, called embryonic layers, is represented. It is from one of these three "stem-like" cell layers that every single cell in your body is derived.[104] These three layers are neatly represented as the center, middle, and outer regions of the outer ear. This gives us unprecedented access to your nervous and immune systems, which communicate with every single cell in your body.

---

103 Scientists call these singularity points in the skin's surface bioelectric field.

104 Nogier P, 1968, 1972, 1983; Nogier P., Petitjean F., Mallard A., 1987, 1989.

## The Nerve that lowers our stress level

Your ear is richly supplied by five nerves that carry both sensory information[105] and motor (muscle) information[106] throughout your entire body, and provides unique access to your nervous system. *Exclusive only to your ear is the access* to what is called the *vagus nerve,* the nerve that regulates the activity of all your vital organs[107] and prevents inflammation. *The vagus nerve is the most important and influential nerve in your entire nervous system for calming you down when you are under stress. It reduces your heart rate and blood pressure as well as changes the function of certain parts of your brain.* It's also important in the *strengthening of your memory* by releasing a hormone (norepinepherine), which helps regulate the learning and the expressing of emotions, and in consolidating your memories. New research has revealed that it may also be the missing link to treating chronic inflammation.[108] Your "stomach" also uses the vagus nerve like a walkie-talkie to tell your brain how you're feeling. Your "gut feelings" are thus very real.

## The number 40

Just as historical writings and artifacts (including historical monuments) make it clear that a cycle of forty days was once used by our ancestors to carefully track time, it's also true that on the fortieth day of human development, our *ear* is formed and becomes enriched with cells that contain the blue-prints for every single cell in our body. The number forty seems to be an important number for humanity, and it marks significant events in ancient writings just as it does for human beings today; after all, it is on the fortieth week that human babies are at full term.

---

105 Such as touch and pressure.

106 For skeletal muscle movement and organ function.

107 Such as controlling your heart rate, breathing, and digestion.

108 V.A. Pavlov et.al, Nat Rev Endocrinol. 2012 Dec; 8(12): 743–754.

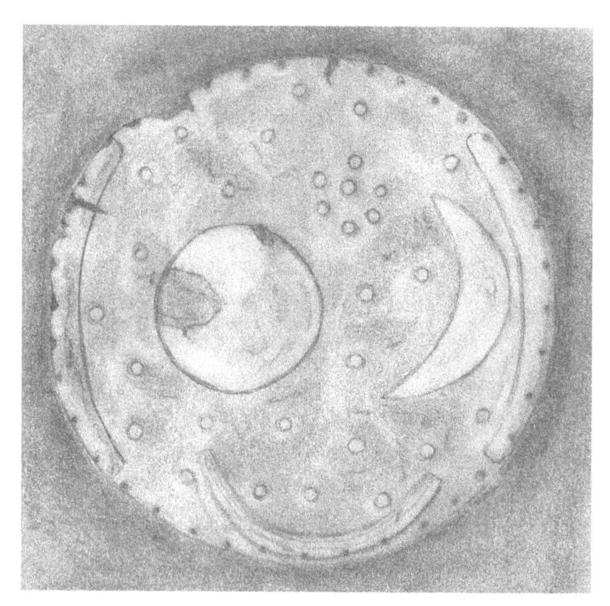

The Nebra sky disk from 1600 BC, with forty perforations along its rim to track time.

## The origins of electrically charged treatment points on the ear

Dr. Nogier, who looked at the individual with an eye of a physicist, also discovered that when the human body is stressed, its electromagnetic field extends out farther than when healthy. An electromagnetic field emits both electricity and magnetism, which will either attract or repel other magnetic objects around it. Our body's *electromagnetic field is projected* from one or several areas on the *outer ear* and can be *detected as changes in electrical flow.* As we first develop, groups of cells that scientists call pluripotent cells (cells that contain information about our entire body and can develop into any type of cell) will gather together to form what are called *organizing centers.* These organizing centers, which are electrically charged, are the areas from which all of our cells form and *they have the same qualities that the electrically active points on our ears have.* We can stimulate our entire body to repair and maintain itself naturally, simply by stimulating these electrically active points on our ear. What a brilliant design.

# Chapter 16

## It's Better to Change an Opinion Than to Persist in a Wrong One (Why America has yet to embrace auricular therapy)

Following a seminar that I attended on auricular medicine, a neurologist came over to me and asked, "Why do so many healthcare providers in the United States still consider the outer ear to be nothing more than a leftover remnant from evolution?" Even though most of us have been taught this, and it's what evolutionists would like us to believe, there's too much evidence today to overlook its value in managing pain and other diseases. I talked with her and others about the reasons why practitioners here in America are so reluctant to use auricular therapy, and in this chapter, I'll share with you the answers.

First of all, even though the World Health Organization published a comprehensive report back in 2003 that included the results of 255 trials and *proved* that acupuncture therapy alone can be an effective treatment for many diseases such as headaches, shoulder pain, lower back pain, arthritis and facial pain, the *claim* here in America is that science cannot precisely explain how it works. And because of this, most American trained providers remain anchored in their ways and dismiss what has been proven and what most of the world acknowledges. In fact, *there is more known about the biologic mechanisms of pain relief by auricular therapy and acupuncture than many procedures of conventional western medicine.* Modern biomedical techniques, including molecular biology and medical imaging, have revealed several correlates (related things) of how auricular therapy works, and, clinical researchers throughout the world have tested and proven its effectiveness. As a result, in 1997, the U.S. National Institutes of Health concluded, ". . . there is sufficient *evidence* of acupuncture's value to expand its use into conventional medicine and to encourage further studies on its physiology and clinical value."[109] A more recent meta-analysis, which combined the findings of twenty-nine randomized controlled studies with a total of 17,922 patients, further demonstrated that acupuncture alone is effective for the treatment of chronic pain,[110] and as previously mentioned, acupuncture accounts for only 10–15 percent of all treatment points on the ear; the remaining 85–90 percent of points on the ear are much more responsive to either pulsed light or very weak electrical stimulation.

The second reason why auricular therapy has gone so unnoticed here in America is due to our history, our relationships with the Far-East and with England, and our belief that we know more than anyone else (i.e., American exceptionalism). Beginning back in 1601, the English East India Company was founded, and its purpose was to

[109] JAMA 280:1518–24.

[110] A. J. Vickers, et al, JAMA Internal Medicine, 2012.

monopolize the East for trading. The English went to India to trade and rule, but not to settle. European competitors, such as the Dutch East India Company, were trading with China and Japan and brought Chinese acupuncture procedures back to Europe. However, by 1767, all Indian Territory was under the sovereignty of the British Crown and as the world power, they discarded any Chinese influences in medicine and culture, and inserted their own. Thus, the benefits of thousands of years of acupuncture medicine and experience were cast away.

Although ear acupuncture was used during the Civil War, and there are published articles from 1826 by Bache Franklin (Benjamin Franklin's great-grandson), it wasn't until 1972, when former President Nixon visited Mao Zedong in Beijing, China, that acupuncture was reintroduced in the United States and was again recognized as being a useful medical treatment. Upon his return, the president's physician, Major General Tkach wrote an article in Readers Digest entitled, *I watched Acupuncture Work*, which helped popularize acupuncture in America. Yet, nearly forty-five years after President Nixon's visit to China, sixty years after Dr. Paul Nogier published *The Treatise of Auriculotherapy* in France, and over three thousand five hundred years since the oldest recorded medical documents first described ear treatments, America has yet to embrace this treasure.

Raphael Nogier shared with me something that his father Paul Nogier personally shared with him. And it's something that my father also impressed upon me; that it's better to change an opinion than to persist in a wrong one.

Chapter 17

# Even Though Ear Points Move, We Will Find Them

In previous chapters, we talked about the different methods that a doctor can use to locate treatment points on the ear, and since only 10–15 percent of all ear points are sensitive to touch, this means that the remaining 85–90 percent of points can only be found by electrical detection. The only proven way to do this is with an instrument that uses two different electrical sensors at the same time. These instruments are called bipolar electrodes (see illustration below). In the United States, the only FDA approved device that can do this is the Stim Flex 400A, which is manufactured in Tulsa, Oklahoma. Dr. Jim Shores, who met with Dr. Paul Nogier numerous times and has close to fifty years of clinical experience, is a well-known expert on its use, and I'm thankful to Jim for sharing his expertise with me.

*Bipolar detection is the only scientifically proven way to locate an electrically active point on the ear* and this is a truth that cannot be overstated. Unfortunately, there are several devices available and in use today that do *not* use bipolar sensors. Such devices are, at best, unreliable.

Detection of electrically active ear points using a bipolar electrode

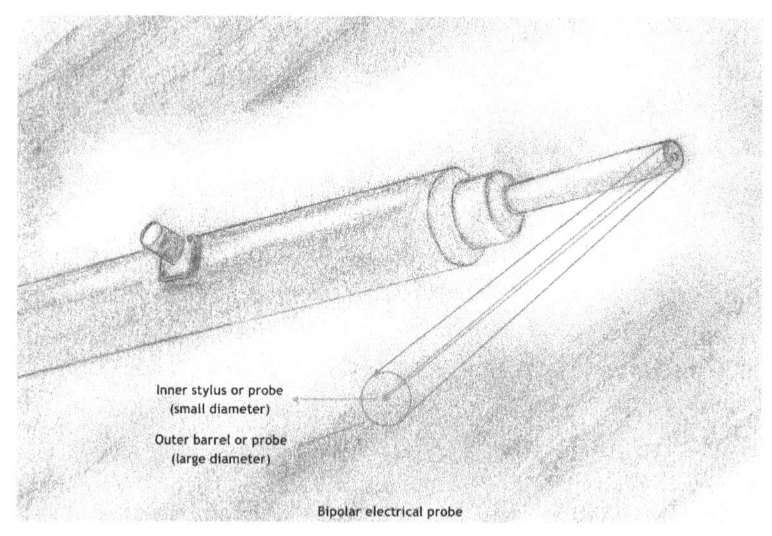


Bipolar (inner and outer sensor) probe

## Everything's in motion

Paul Nogier was known to be a very private person. But his passion for healing others and his relentless work ethic were widely recognized. Rather than resting on his past works, he continued his research and looked for ways to improve and refine his discoveries. As a result, his treatment methods and theories would often evolve to reflect his new findings. Thankfully, he wasn't afraid to change his opinions. But change can be uncomfortable, especially when it's substantial.

Originally, in 1969, his discoveries led to the publishing of a map of treatment points on the ear. These points were fixed in one location and didn't move. But as the result of his continued research, he revised his original map to show that the ear points do move from one area to another. He concluded that the surface of the ear is a holographic, three-dimensional, energetic projection of the human body. This meant that every point that was meticulously mapped out on his 1957 ear chart could now move to two additional locations. This sent shockwaves through the scientific community. This complicated things. But the explanation of why the points would move was actually simple. Dr. Nogier discovered that their location was dependent upon whether a person's condition was acute (sudden onset), degenerative (causing loss of function), or chronic (long-lasting). Regardless of where on the ear the points move to, we can find them readily with bipolar detection.

Very few doctors have taken auricular therapy to the depth of Nogier's complex and sophisticated three-phase approach. Most practitioners, including those using the Chinese system, utilize only the one-phase approach that was first introduced in the 1950s. In my practice, I let the ear be my guide rather than being limited by an ear chart. Whether it's by electrical activity, or by being sensitive to the touch, I let the ear guide me to the location and to the type of treatment that's needed.

## What does a change in the strength of our pulse mean?

Nearly two thousand years ago, Galen, known as the greatest physician of the Roman Empire, recognized the importance of the pulse. More recently, Paul Nogier discovered a unique ear-heart-reflex that also emphasized the value of the pulse. Today, thousands of doctors around the world utilize this reflex to help with their diagnosis and treatment, and its potential is yet to be fully realized.

As your heart beats, it doesn't provide enough force to carry blood to the most distant areas of your body and through the smallest arteries called arterioles. The good news is that both your arteries and arterioles are lined with small muscles that contract and relax to complement the work of your heart and to ensure that your entire body receives the nutrients it needs. When your body is stimulated, it responds with an instantaneous change in the tone of the wall of all the arteries and this reflex is called the VAS or Nogier Vascular Autonomic Signal.

So what does a change in the strength of your pulse mean? It means that your body is responding to information that it's receiving, and it's adapting to it. This information can be in the form of light, or physical contact such as message, or from an object that's held close to but doesn't come in contact with the skin (i.e., allergy testing). This response is like a "readout" of your body's sensitivity to any subtle changes and it's used to identify any imbalances, as well as to provide information on how best to manage them. What I am feeling is your body unconsciously adapting to a stimulus and working to return to a balanced state. In short, it's like a guide that gives me an update on how healthy and balanced you are.

So what does the Nogier pulse feel like? It feels like a tire that's being pumped up when the pulse strength increases and it feels like a tire that's being deflated when the pulse strength decreases, which happens during allergy testing when materials that a person may be allergic or hypersensitive to are brought near the skin.

While all of our skin is saturated with tiny acupuncture points that act like antennae to both receive and transmit information from the outside and from within, the skin that covers our ear is especially unique. This is because unlike any other area on our body, it only becomes electrically active when something is wrong and there's an imbalance. This electrical activity is actually information that tells us where in our body there's an imbalance and what kind of treatment is best.

## How Dr. Nogier discovered the pulse reflex

In 1966, Dr. Nogier, by chance, recognized an extraordinary phenomenon. While taking the pulse of a patient and simultaneously exerting pressure on the ear with a small ballpoint pen, he observed that the radial (wrist) pulse changed in strength and became weak and distant. Meanwhile, the patient's heart rate remained unchanged. He repeated this experiment hundreds of times over the course of thirty years and studied the reactions of the radial pulse while the ear was being stimulated by touch, temperature, light, electrical frequencies, and vibration. Dr. Nogier went on to discover that stimulating the skin anywhere on the body could also trigger this phenomenon.

Whether our skin is touched with a cotton ball or stimulated by a flash of light, our arteries will react and this can be felt as a change in our pulse strength. Dr. Nogier measured the reaction of the body to a stimulus by counting how many times the pulse became stronger or weaker before it returned too normal. This observation corroborated with the 1917 and 1943 findings that were published by a French vascular surgeon Professor Rene Leriche. Dr. Paul Nogier, the father of auricular therapy, trained his son, Dr. Raphael Nogier, to recognize the subtleties of this reflex and Raphael graciously shared his knowledge with me.

Dr. Kalfus and Dr. Nogier

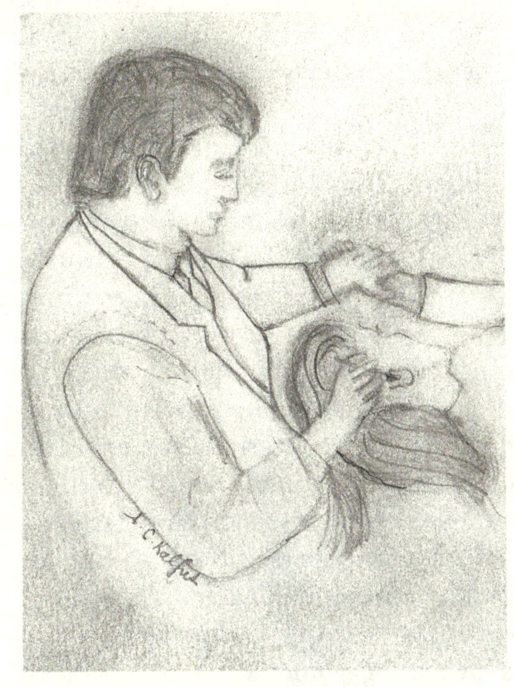

Nogier pulse while evaluating ear

## Obsession

Our nervous system is obsessed with maintaining a balance within our body. This is accomplished by the existence of a very complex control system. We know about our sensors (such as our eyes, ears, nose, mouth, and skin), which act like antennae to provide us with conscious information. However, we don't really know much about the sensors that are distributed throughout our body that provide unconscious information to our nervous system.

When our senses gather information, our nervous system processes it with the aim of maintaining or restoring a balance of our internal environment. Our nervous system has a memory of past events that influences this response. The response will either be a conscious action or an unconscious action.

To initiate a conscious action (one that we're aware of), our nervous system sends us a signal that is perceived as either pleasant or unpleasant. Our body will do whatever it can to normalize or balance its internal environment. For example, when we are hungry, our nervous system (specifically the hypothalamus of our brain) will trigger a signal for us to eat in order to soothe this unpleasant sensation. Another example is when we become dehydrated and the concentration of sodium (salt) in our body rises, our nervous system will trigger a stimulus perceived as thirst and we will seek water to reduce this unpleasant sensation.

To initiate an unconscious reaction (one that we're not aware of), our nervous system will modify the amount of work that our organs have to do, such as making our heart or liver work harder, which changes the number of calories that are burned. This change will immediately impact the balance within our body. *The unconscious reactions will act mainly on the smooth muscles* of our organs (i.e., digestive tract, muscles, and arteries), our endocrine glands (i.e., thyroid and adrenal glands) and also on our immune system. The speed of onset of these reactions is virtually *instantaneous* and can be seen for example, when a person turns pale or red when they're frightened. This is an adaptation to a new situation.

Among the consequences of this unconscious response to the environment is the *tensing of muscle cells in the walls of arteries.* The tensing of the arterial walls occurs because the walls are sensitive to substances such as adrenaline that are released into the bloodstream when we are trying to *adapt* to a new situation. *The change in the strength of our pulse is actually an expression of our nervous system as it quickly adapts to a change in the environment and it seeks to maintain or restore internal balance.* It's a phenomenon that allows a practitioner to evaluate the adaptability of a patient.

## The rabbit test

When it comes to the topic of light therapy on the ear, it's good to know what really works and what doesn't. Some doctors insist on using colored light, such as blue, green, or red light, while others say the color of the light has little to do with the success of treatment and it's the frequency at which the light pulses on and off that's key. I spoke about this very subject in length with Raphael Nogier and he shared with me the results of a well-known experiment that he personally conducted in 1980. He wanted to know if exposing the skin of rabbits with light would change the blood levels of certain hormones that are known to contract arteries. There are only three natural hormones that do this: dopamine, epinephrine and norepinephrine. Dopamine is associated with pleasure and motivation, as well as with memory, attention, and problem-solving. Epinephrine, also known as adrenaline, causes an increase in heart rate, muscle strength, blood pressure, and sugar metabolism for quick energy. Norepineperine is released in response to stress, and it works hand in hand with adrenaline to increase our heart rate, open up airways, and do other things to prepare our body for a perceived threat or stressful situation. Dr. Nogier's experiment was simple. Rabbits were placed in a cage with a hole on one side. The rabbits had their heads out of the cage and could not see what was happening inside. Three groups of rabbits were studied. The first group was exposed to a single continuous white light, the second with white pulse light at 9 Hertz (i.e., the light flickers on and off nine times every second), and the third group was the control group with no light. Both the first and second groups received the same amount of photons (light) on the skin. *The experiments revealed that blood hormone levels increased significantly only for the rabbits exposed to the intermittent light.* This group also showed an increase in blood glucose (sugar) levels. In contrast, the continuous light did *not* show any increase in the blood levels of hormones that cause arteries to contract. In fact, there was an increase in a substance (called cholinesterase) that actually breaks down hormones that lead to muscle contraction (i.e., helps with muscle relaxation).

The rabbit test showed that the skin is capable of discriminating between certain messages that are within the light (known as electromagnetic messages). This phenomenon is referred to as cutaneous (skin) photoreception, which is the skin's ability to receive and transmit information that's within light. Interestingly, this phenomenon was briefly studied in 1937 by a well-known French physiologist named Jules Tinel. He wondered whether the skin was a receiver of unconscious information. We now know that *the skin contains many free nerve endings*[111] *as well as specialized immune cells* (called Langerhans cells) *and both are known to be very sensitive to light.* We also know that *electrically active ear points all have a small artery, vein, free nerve endings, and a lymphatic duct* (which returns fluids back into the blood) *with Langerhans cells.*[112] Thus, *electrically active points on the ear are very responsive to light frequencies. And*

111 Nerves that sense pain, pleasure, and temperature.

112 Immune cells that help fight against infection.

*by sending a flash of white light on the skin, we will induce an arterial reaction,* which can be detected by feeling a change in the strength of the pulse.

## A simple and comfortable way to test for allergies

Our skin seems to recognize different chemicals that we're exposed to by sensing changes in the light as it passes through a chemical (i.e., allergen) and then shines onto our skin. Food and allergy tests can be conducted by bringing potential allergens near the skin of the patient while observing the reactions of the pulse. As light passes through a sample of potentially allergic material,[113] the light will pick up information about what's in the sample and the skin will recognize these materials.

We know that the cells of our nervous system, called neurons, communicate using chemical messages. These cells do not touch each other. We also know that our skin and the nervous system have very similar structures and both are derived from the same embryological layer known as the ectoderm. Our skin appears to be like a telescope that captures electromagnetic information that surrounds us (including sound, light, X-ray, and gamma ray information) and is able to both receive and understand this information, which can be used to help us to regain and maintain our health.

---

[113] That is wheat, gluten, dairy samples that are placed in a thin prepared microscope slide.

# Chapter 18

# Our Ear Is Intimately Linked to Every Cell in Our Body

The biological life of a human being begins as a union of cells that divide and multiply, forming a ball of tissue with increasingly more individual cells. These cells continue to multiply until they ultimately become an embryo that has three concentric (objects that share the same center) layers. Each of these layers, from which every single cell in your body is formed, is represented in the outer ear.

Next to genetic imprinting, a growth control system directs cellular growth and differentiation. This system is the first physical communication system in an embryo. Cells mainly grow in the direction of a negative charge and groups of cells are called *organizing centers.* These *organizing centers communicate with each other by channels that extend across gaps between the cell groups.* The basic idea is that the *ear points that emit electricity and are used in auricular therapy are remnants of this growth control system which, along with a Traditional Chinese Medicine meridian system,*[114] *were intimately linked together during development, and they continue to stay connected and communicate with all of our body systems.* Therefore, our ear gives us unique access to real-time updates on our health and provides for repair and maintenance unlike any other area of our body.

---

114 A traditional Chinese Medicine belief that acts like a giant web linking all of our systems together.

# Chapter 19

## The Science behind Auricular Therapy

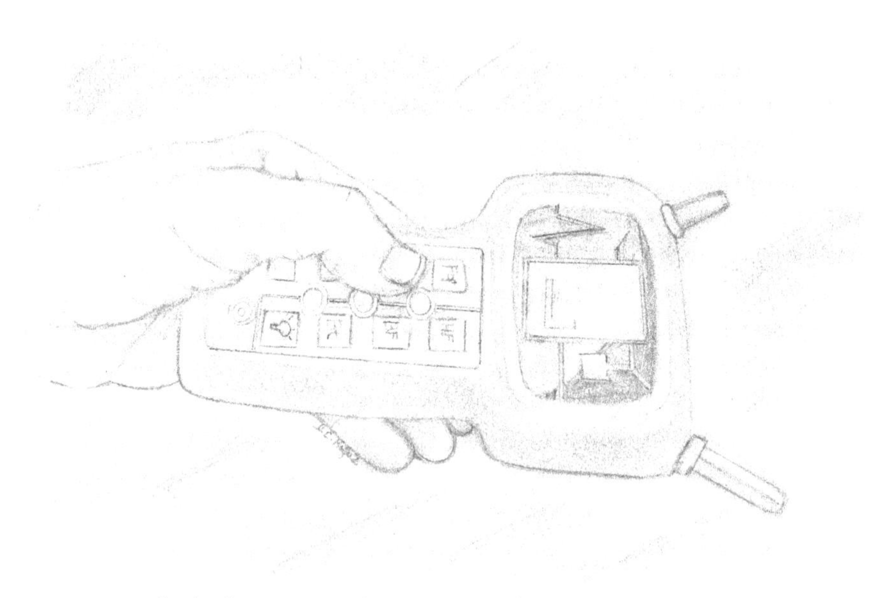

Light frequency detection and treatment device

Anytime something looks different from what we are used to, we tend to give it a lot of attention. Sometimes this is good, and sometimes it's not. It's natural for us to feel comfortable with things that we're familiar with, and to feel cautious or even threatened by things that are foreign to us. Auricular medicine is certainly no exception to this rule.

Each of us should have a sense of comfort and confidence when it comes to treatment that we're either about to receive or provide. Here are the questions that I'm most frequently asked, and it's my hope that the answers will satisfy any concerns that some of you may still have.

- What's the scientific evidence that supports its use and how does it actually relieve pain and speed up healing?
- What conditions can it be used for?
- Can it make matters worse?

A colleague of mine, whenever he's asked similar questions will usually respond by saying, "I've been wrong once in my life. I thought I was wrong twice, but I was wrong." I smile every time I think about him saying that.

The remainder of the book is for those of you who want to know the science, the evidence, and the mechanisms that support and validate this remarkable treatment. The following topics are included.

- The physiological basis for auricular therapy.
- What conditions has auricular therapy been proven to benefit?
- Is chronic pain due to an alteration of natural healing neural circuits?
- How auricular therapy affects the neurophysiology of pain.
- Bioelectromagnetic radiation and Biophotons—All living things emit light.
- The liquid-crystalline form of water and its importance within the core of our DNA for electromagnetic reception and transmission.
- The energy that comes from the sunlight is the key ingredient to hexagonal water molecules found in the core of our DNA.
- Hexagonal Gap—junctions between cells help with communication.
- How auricular therapy relieves pain.
- Histological features of neurovascular (electrically active) ear points.
- Unique physiological features of neurovascular ear points.
- Stimulation of the ear directly effects brain activity/ neuroimaging.

- Identifying endorphin and neurochemical correlates to auricular therapy.
- Bidirectional effects of therapy.
- The ear as a peripheral microcosm of the entire body.
- Embryological derivatives on the ear.
- Two-way reflexes of the ear. Information is both taken and received.
- How visual changes on the ear correlate to disease.

Although the science that supports auricular medicine can seem complicated, its application is simple. Whether it's a gentle pulse of light, the application of current at millionths of an amp, or the use of tiny needles on the ear, auricular therapy has proven itself to be a safe, fast and effective way to relieve the pain and functional problems associated with many medical complaints. It's my hope that it will be readily available to anyone who wants to live a more pain-free life, accelerate their natural healing process, and restore and maintain their health.

# Chapter 20

## Physiological Basis for Auricular Therapy

For years, the pain-relieving effects of acupuncture have been explained away by the well-known placebo effect, which works through suggestion, distraction, or even hypnosis. But how does this explain the use of acupuncture analgesia[115] in Veterinary Medicine over the past thousand years in China and one hundred years in Europe and its growing use on animals in America?

Animals are not suggestible and only a few species are capable of the still reaction (so-called "animal hypnosis"). Similarly, small children also respond to acupuncture anesthesia. Several studies in which patients were given psychological tests for susceptibility did not show a good correlation between acupuncture analgesia and suggestibility.[116] Hypnosis has also been ruled out as an explanation by studies[117] that have shown that hypnosis and acupuncture analgesia respond to naloxone differently; *acupuncture anesthesia being blocked and hypnosis being unaffected* by this endorphin antagonist. Naloxone is a drug used to reverse the effects of a class of highly addictive drugs called Opioids (like Vicodin and Oxycontin) that are commonly prescribed for pain relief. Naloxone only works when opioids are present and it has a stronger affinity to opioid receptors then opioids. Auricular therapy stimulates the release of natural opioids called

---

115 The loss in ability to feel pain due to acupuncture.

116 (Liao SJ, 1978).

117 (Barber J, Mayer DJ, 1977; Goldstein A, Hilgard EF, 1975)

endorphins and when Naloxone is introduced, it takes the place of the endorphins on the opioid receptors and prevents the receptors from being stimulated. Without opioid receptor stimulation, there is no analgesic effect. Thus, there is a physiological basis to Auricular acupuncture.

Decades of research have clearly shown that the tiny needles used in acupuncture stimulate small nerve endings, which send impulses to the central nervous system (CNS) and activate the release of endorphins, enkephalins, and dynorphines, often called nature's pain relievers or endogenous morphines, and monoamines (dopamine, noradrenalin, and serotonin), which are also neurotransmitters. Auricular Acupuncture has been found to raise blood serum and cerebrospinal fluid levels of endorphins and enkephalins.[118] Supportive findings in back pain patients were obtained by Clement-Jones et al.[119] where low-frequency electrical stimulation of the concha led to the relief of pain within twenty minutes of the onset of stimulation and an accompanying elevation of CSF beta-endorphin activity was demonstrated in all subjects. In 2001, Pomeranz reviewed the extensive research on endorphins and acupuncture and convincingly established that endorphins have a scientifically verifiable role in explaining the pain-relieving effects of acupuncture. Eriksson in 1976 and Abbate et al., in 1980 demonstrated a significant increase in beta-endorphins after auriculoltherapy. Along with the pituitary release of endorphins is the important co-release of Adrenocorticotropic Hormone (ACTH), which stimulates the adrenals to raise blood cortisol levels after acupuncture analgesia. Sham acupuncture does not raise cortisol levels, thus ruling out stress as the cause of cortisol release. In addition to changes in plasma and CSF levels of endorphins, other biochemical changes have been proven to accompany auriculotherapy. Debreceni (1991) demonstrated an increase in plasma growth hormone and Jaung-Geng et al. (1995) showed that auricular acupressure at the appropriate acupoints significantly reduced the toxic buildup of lactic acid due to improved peripheral blood circulation.

---

[118] Pomeranz and Chiu, 1976; Sjolund et al., 1977, 1996; Clemente-Jones et al., 1979, 1980; Ho and Wen, 1989; Chen, 1993; Gao et al., 2008, 2011, 2012.

[119] Clement-Jones V, McLoughlin L, Tomlin S, Besser G, Rees L, Wen H, 1980.

As mentioned above, the tiny needles used in auricular acupuncture induce small changes in the skin that activate the survival mechanisms that normalize homeostasis and promote healing.[120] The needling is perceived as a foreign invader and the needle-induced "lesion" as an injury due to mechanical trauma to the surrounding tissues. It may cause tiny bleedings in the soft tissues and thus induce hemodynamic changes such as dilation of the arterioles and the opening of new capillaries and venular beds in the area. When needles are inserted directly into injured and inflamed areas, a small local anti-inflammatory reaction is triggered that speeds up the inflammatory process in an effort to protect your body. After the acupuncture needle is removed, the needle-induced lesion continues to stimulate your body until the lesion heals.[121] Needling also reduces bodily stress by stimulating the parasympathetic nervous system, thereby relaxing the cardiovascular and muscular systems, and restoring the physiologic and autonomic balance that includes normalizing visceral functions that were impaired during the stress of acupuncture. Centrally, the small changes in the skin's surface and surrounding tissues stimulate the survival systems in the brain such as the nervous, endocrine, immune, and cardiovascular systems and work to normalize the physiological activities of your entire body. Auricular therapy can be thought of as a physiological therapy whereby the brain responds to the stimulation of peripheral sensory nerves that are proximal to the ear point being treated. The therapy does not treat a particular symptom, but it normalizes the physiological balance and promotes self-healing without side effects and treats both the cause and the symptoms of the ailment; it is a physiological regulator.

[120] Sandberg M., et al., 2005; Shinbara H., et al., 2013.

[121] Langevin, 2001; Langevin, 2007.

# Chapter 21

## What Conditions Has Auricular Therapy Been Proven to Benefit?

Acupuncture has been proven to be beneficial in a variety of painful conditions to include chronic neck pain,[122] low back pain,[123] osteoarthritis,[124] migraines,[125] symptoms associated with fibromyalgia,[126] tension headaches,[127] palliative care of cancer to alleviate pain, dysfunction and fatigue,[128] peripheral neuropathy[129] and has been shown to be superior to physical therapy and usual care in patients with chronic low back pain.[130] For migraines, acupuncture has been shown to be more effective, provide more immediate and long-lasting relief, and have no adverse side effects as compared to drug therapy. Auricular therapy has also been shown to be an effective

---

122 White, P., et al. 2004.

123 Molsberger A, 1998.

124 Berman, B.M., et al. 2004; Scharf, H.P., et al. 2006; Witt, C.M., et al. 2006.

125 Facco, E., et al. 2013; Vickers Aj, Cronin AM, Maschino AC, et al. 2012.

126 Martin, D.P., et al. 2006.

127 Linde K, Allais G, Brinkhaus B, Manheimer E, Vickers A, White AR. 2009 meta-analysis.

128 Alimi, D., et al. 2003; Pfister, D.G., et al. 2010; Vickers, A.J., et al. 2004; Molassiotis, A., et al. 2012.

129 Schroder S, Liepert J, Remppis A, Greten JH. 2007.

130 Hsieh, l. l., et al., 2006; Thomas, K.J., et al. 2006; Haake, M., et al. 2007.

treatment for acute anxiety,[131] insomnia,[132] dental pain[133] to alleviate pain both during and after dental operations; to relieve pain and anxiety in patients with lumbar disc hernia,[134] in prehospital transport phase of hip fracture[135] and to reduce acute pain due to a variety of causes in the emergency department setting.[136] In summary, both an individual patient data meta-analysis on acupuncture for chronic pain[137] and a systematic review and meta-analysis that included 17 studies on acute pain, chronic pain, and postoperative pain showed that auricular therapy was an effective treatment for a variety of types of pain and is a good adjunct for patients having difficulties with pharmacologic pain therapies and medications.[138]

The Chinese have an old saying: "That nine of ten diseases produce pain". And according to statistics, 85 percent of pain is soft tissue pain.[139] Fortunately, it is now a well-established fact that auricular therapy is effective for a variety of health problems, and it is especially effective in the management of pain.

---

131 Sarst M, Winterhalter M, Munte S, Franckl B, Hondronikos A, Eckardt A, Hoy L, Buhck H, Bernateck M, Fink M, 2007.

132 Sok SR, Kim KB, 2005.

133 Ernst E, Pittler MH, 1998.

134 Liu EJ, Jia CS, Li XS, Shi J. Zhongguo Zhen Jiu. 2010.

135 Usichenko, T.I., et al. 2005; Baker, R., et al. 2006.

136 Goertx, C.M., et al. 2006.

137 Vickers, A.J., et al. JAMA Intern Med. 2012.

138 Asher, G.N., et al. 2010.

139 White, A., Cummings, M. and Filshie, J. 2008.

# Chapter 22

## Is Chronic Pain Due to An Alteration of Natural Healing Neural Circuits?

There are a number of explanations or theories to account for the connections between painful and/or active neurovascular/electrical points on the skin and the brain. Tsun-nun Lee, in 1994, stated that pathological changes in peripheral tissue could lead to maladaptive, dysfunctional firing patterns in the connections between peripheral nerves and the CNS. These connections are organized and controlled by sites in the sensory thalamus. But strong environmental and/or emotional stressors can alter these natural healing circuits to a point where they become impaired (they misfire) and new dysfunctional patterns are learned. All this leads to pain and disease.[140] Stimulation of correspondent points on the ear sends messages to the brain to promote a healthy reorganization of these circuits to return them to their natural patterns.

---

[140] Nogier, 1983.

# Chapter 23

## How Auricular Therapy Affects the Neurophysiology of Pain

Your body emits low-level light, heat, and acoustical energy and has electrical and magnetic properties.[141] These emissions are the basis of a number of conventional medical tests such as the electrocardiogram (ECG) and the electroencephalogram (EEG) that assess the physiologic function of the heart and brain, respectively. The ECG was first developed in 1887 while the EEG was developed in 1875. More recently, *corresponding magnetic field measurements* of the heart and brain have been discovered, which are the magneto cardiogram (MCG) and the (MEG) magneto encephalogram, respectively. However, the *magnetic fields of the body are of very low level* and typically require specialized equipment such as super-conducting quantum interference devices (SQUIDs). And recently, it has been observed that living organisms emit *bioelectromagnetic* (BEM) radiation in the form of *biophotons (Greek bio=life, photon=light)* that contribute to the regulation of a range of cell biological events from membrane transport to gene expression.[142] The detection of biophotons or "low-level luminescence," which will be discussed in more detail later, is facilitated by photomultipliers that greatly amplify the photons emitted so that they can be counted, and this measurable electrical energy is what we tap into with auriculotherapy. Biophoton

141 Pang, X. 2008; Gu, Q. 1989; Popp, F.A., 1989; Popp, F.A., Chang, J.J., & Gu, Q. 1996; Popp, F.A., & Chang, J.J. 2000; Cifra, M., Fields, J.Z., Farhadi, A 2010.

142 Popp, F.A., Nagl, W., Li, K. H., Scholz W., Weingartner O & Wolf, R. 1984.

emission from acupuncture points is generally higher than that of the surrounding skin[143] just as the electrical conductivity at auricular acupuncture points is often higher than the surrounding skin.[144] Also in connection with the biophoton emission and the electrical activity that characterizes auricular reflex points is the close proximity of neurons, the fundamental unit of the nervous system, and one of the few types of biological cells that carry electrical signals. One aspect of neurons that corresponds to the electrodermal features of auricular acupuncture points is a series of low skin resistance gaps (nodes of Ranvier in neurons) separated by regions of high skin resistance nonacupuncture points. The speed of neural impulses is increased by the presence of segments of high electrical resistance myelin separated by gaps of lower electrical resistance at the nodes of Ranvier. Each neuron is capable of rapidly carrying electrical neural impulses and in ear acupuncture, the mechanoreceptors, thermoreceptors, and sensory receptors excited by noxious stimuli, called nociceptors, are located in the superficial skin surface. These receptors all travel along together, like the individual wires in an electrical cord. The initiation of pain signals begins with the activation of these microscopic neural endings found in the skin, the muscles, the joints, the blood vessels, or the viscera. Following injury, there is an array of biochemical's released to include prostaglandins, histamine, bradykinin, and substance P. These chemicals activate your nociceptors, in contrast to the other sensory receptors that are activated by light touch and moderate temperature changes. The thinnest neurons, which have no myelin coating, are called type C fibers and tend to carry information about nociceptive pain. The next larger neurons are called type B fibers, which have a little myelin coating and tend to carry information about skin temperature and internal organ activity. The thickest diameter neurons are call type A fibers and are myelinated and large in size, making them much faster than the type B and type C fibers. Type A fibers are further subdivided into the larger A beta and smaller A delta fibers. The A beta fibers carry information about light touch while the A delta carry information about immediate, nociceptive pain. *When you get*

---

143 Inaba H, 2000; Inaba H, 1998.

144 Nakatina, Y. 1950 in Japan; Niboyet, J. 1963 in France; Yiu-Ming Wong, 2014.

*hurt, there is an immediate "sharp" response by A delta fibers followed by the delayed "throbbing, enduring" perception of a second pain that is carried by the slower C fibers.* Chronic illness, in order to maintain itself may be due to an alteration in the neurological reflexes that transmit pathological messages to the higher nervous centers.[145]

In the 1970s, neuroscientists discovered that the brain not only contains certain areas that can sense painful signals, but it also contains descending pain inhibitory pathways.[146] Examination of deep brain stimulation in people has shown that while the midbrain periaquiductal grey (PAG) plays a major role in analgesia, *the thalamus is the most potent site to yield stimulation-produced analgesia.*[147] Research has confirmed that nociceptive pain messages activate positron emission tomography (PET) scan activity in the periaquiductal grey (PAG), thalamus, hypothalamus, somatosensory cortex, and prefrontal cortex.[148] These are the same brainstem and thalamic areas that are able to suppress pain messages. Neurophysiologic investigations demonstrate that these regions are also affected by stimulation of acupoints[149] such as those mapped out on the external ear. So this descending neural inhibitory pathway that can inhibit the ascending pain message before it gets to the cerebral cortex for interpretation includes the cingulated gyrus, the amygdala, the thalamus, the midbrain periaqueductal gray, and the reticular formation within the brainstem. Interneuron's within the dorsal horn of the spinal column have the ability to turn off the pain-related neurons and are thus called pain-gating neurons. Again, the external ear has specific corresponding acupoints for each of these regions to inhibit the ascending pain message. Note that lesions of the PAG nullify acupuncture analgesia indicating that the anatomical integrity of the PAG is necessary for producing pain relief from auriculotherapy. Afferent signals project from the PAG to the posterior hypothalamus, the lateral hypothalamus, and the central, median nucleus of the thalamus. These neurons project to

---

145 Nogier, 1983; Tsun-Nin Lee 1994.

146 Liebeskind et al., 1974; Mayer et al., 1971; Mayer and Liebeskind, 1974.

147 Hosobuchi et al., 1979.

148 Hseih et al., 1995.

149 Kho and Robertson, 1997.

the hypothalamic pre-optic area to the pituitary gland, from which beta-endorphins are secreted into the blood to provide pain relief.

In addition to the evidence provided by PET, serum, and cerebrospinal fluid changes that directly relate to auriculotherapy, fMRI studies of the human brain (Cho and Wong, 1998; Cho et al., 2001) provide further direct evidence of the neurological effect of acupuncture stimulation. Dr. David Alimi of France specifically focused on auricular stimulation.[150] fMRI of the human brain by Dr. Z. H. Cho show much higher preacupuncture brain activity in the prefrontal cortex, cingulate gyrus, the thalamus, and the periaqueductal gray than the fMRI images taken after acupuncture stimulation. This suggests that the acupuncture stimulation led to a reduced response to painful stimulation. All of these changes point to an unconventional energy that profoundly affects the healing benefits of auricular therapy. Dr. Paul Nogier discovered that the electromagnetic field of an individual underwent a measurable change when under stress. A fraction of a person's bioelectromagnetic field is emitted in the form of a measurable unit of light, known as a biophoton. *The biophoton emission at electrically active ear points is altered by auricular therapy, which initiates a sequence of events that promotes not only dramatic changes in the nervous system but also a reorganization of circuits back to a balanced, healthy state.*

---

150 Alimi, 2000; Alimi et al., 2002.

# Chapter 24

## All Living Cells Emit Light

As previously mentioned, biophotons, or ultra-weak photon emissions of biological systems, are weak electromagnetic waves in the optical range of the spectrum—or simply put, light. A photon is a single quantum of visible light, a small discrete package of electromagnetic radiation some of which is light. This light emission is an expression of the functional state of the living organism and its measurement therefore can be used to assess this state. Cancer cells and healthy cells of the same type, for instance, can be discriminated by typical differences in biophoton emission. *All living cells emit biophotons that cannot be seen by the naked eye* (they are a thousand times less intense than what your eyes are sensitive to) but can be detected by specialized equipment (photomultiplier tube) that was first introduced in the 1950s and used by Strehler and Arnold. This device can count the number of individual photons produced by living cells. *The main source of this light energy arises from an organelle found in every cell of the body called the mitochondria.* The mitochondria are known as the "cellular power plants" because they convert organic materials into energy in the form of Adenosine Tri-Phosphate (ATP) via the process of oxidative phosphorylation. There can be hundreds or thousands of mitochondria in every cell (can occupy 25 percent of cytoplasm and have their own DNA). *There are 100,000 chemical reactions per cell/per second, all of which must be carefully timed and sequenced with each other. Without electronic excitation of at least one of the reaction partners within the cell, it would be impossible for it to have this huge amount of chemical reactivity.* So even when you rest, you are swimming in a sea of light, part of a sophisticated dance that has you buzzing with activity. Free radicals (reactive oxygen species such as superoxide and hydrogen peroxide) are produced inside the mitochondria and are associated with cell damage and may be responsible for the emission of biophotons. A small percentage of electrons (1–4 percent) in the electron transport chain leak onto an oxygen molecule. This oxygen molecule is called *superoxide* but is unstable because it needs an extra electron on its outer shell. Superoxide is prone to steal an electron from the nearest source such as the mitochondrial DNA, the mitochondrial membrane, proteins, or vitamins C and E. Borrowing electrons from vitamins or nearby antioxidants does no harm to the cell. *This is why you would want to eat your fruits and vegetables because they contain antioxidants, which lend electrons to the superoxide molecule and won't need to borrow from structures such as mitochondrial DNA. Otherwise, cell damage will lead to cell death, and this isn't good for you. There are proteins on the mitochondria that detect damage, which in turn activate other proteins to punch holes in the mitochondrial membrane and this eventually leads to cell death. Aging occurs as mitochondria become less functional or die out.* As the cell can no longer function, aging accelerates. This is why a diet rich in antioxidant is a good idea. It's also interesting to note that *as stress levels increase, so do biophoton emissions.*

I am always careful with the term "ultra-weak" biophoton emission because it can be misleading since it suggests that biophotons are not too important for cellular processes. Estimates indicate that for a measured intensity of biophotons, the corresponding intensity of the light field within the organism can be up to a hundred times higher.[151] According to Bokkon et al.,[152] the real biophoton intensity within cells and neurons can be considerably higher than one would expect from the measurement of ultra-weak bioluminescence, which is generally carried out macroscopically several centimeters away from the tissue. In other words, the most significant fraction of natural biophoton intensity cannot be accurately measured because cellular processes absorb it. Biophotons can be absorbed by natural chromophores, which are the parts of a molecule that absorb certain wavelengths of visible light giving it its color.[153] *The absorption of biophotons by a photosensitive molecule can produce an electronically excited state.* As a result, molecules in electronically excited states often have very different chemical and physical properties compared to their electronic ground states. The biophotons are absorbed by nearby chromophores and can excite nearby molecules and trigger or regulate complex signal processes. Thus, the absorbed biophotons could have effects on the electrical activity of cells and neurons via signal processes. This subtle yet complex cell-to-cell communication relies on the speed of light transmission through the air rather than the slower communications that occur by physical contact between cells, through fluids such as blood (hormones) or interstitial (between tissue) fluid. So in a *way*, it's *OK* to *say* that living systems "suck" the light *away* in order to establish the most sensitive platform of communication throughout the *day*.

*A significant increase in photon emission is evident around sites of tissue injury and this is an expression of the functional state of a living organism and its measurement, therefore, can be used to assess its health or lack thereof.* In other words, reactions to stress are often indicated by an increase in biophoton emission.[154] Furthermore, there is evidence that the different states of DNA influence biophoton

---

151 Chwirot BW, Popp FA, Li KH, Gu Q, 1992; Slawinski J, 1988.

152 Bokkon I, Salari V, Tuszynski J, Antal I, 2010.

153 Karu T, 1999; Kataoka Y, Cui Y, Yamagata A, Niigaki M, Hirohata T, Oishi N, Watanabe Y, 2001; Mazhul VM, Shcherbin DG, 1999; Thar R, Kuhl M, 2004.

154 Popp F A, Ruch B, Bahr W, Bohm J, Grass P, Grolig G, Rattemeyer M, Schmidt H G & Wulle P-1981; Popp F A, Gu Q & Li K H-1994; Popp F A, Li K H & Gu Q-eds-1992.

emission indicating that chromatin[155] is one of the most essential sources of biophoton emission.[156] *The increased emission of biophotons at active auricular points is measurable and correlates to changes in electrodermal resistance and conductance and thus serves as a signal or alarm that allows us to specifically locate and treat the corresponding part of the body that is injured or imbalanced.*

Additionally, bioelectromagnetic radiation in the form of biophotons contributes to the regulation of a range of cell biological events from membrane transport to gene expression.[157] There is evidence to show that bioelectromagnetic fields arise from the direct current generated by cells that surround neurons[158] and we know that acupoints are surrounded by nerves. Based on this, it is likely that a skillful practitioner using auricular therapy can generate weak electromagnetic fields that modify or "pattern-correct" abnormal bioelectromagnetic fields within your body[159] to regulate energetic, molecular, and cellular events that lead to healing and health.

---

155 A macromolecule found in cells that consists of DNA, RNA, and proteins.

156 Chwirot B-1986; Rattemeyer M, Popp F A & Nagl W-1981.

157 Rubik B, Becker RO, Hazlewood CF, Liboff AR, Walleczek J, 1995.

158 Becker RO, Marino A, 1982.

159 Becker RO, 1982.

# Chapter 25

## The Hub of Life Is Water and Light

As mentioned in chapter 4, the earth is a watery place and on average, human beings are made up of about 75 percent water (at a molecular level, our bodies are more than 99 percent water) and staying hydrated is essential to our health. *Water found in healthy human*

*cells*, however, is not ordinary. Rather, it is *highly organized*. And *the key ingredient to water molecules becoming organized is sunlight* (electromagnetic energy). Among the total spectrum of solar rays coming from the sun, the greatest amount of the sun's energy output is in the infrared segment of the electromagnetic spectrum, which is further divided into the near infrared, middle infrared, and the far infrared rays of light. It is the *far infrared waves that are the safest and the most beneficial*. The world's leading authority on water science, Dr. Mu Shik Jhon, concluded that it is *how water molecules bond together that may be a key to health and aging*. He summarized his forty years of research into a simple theory that defines aging as a loss of hexagonal water from organs, tissues, and cells, and an overall decrease in total body water.

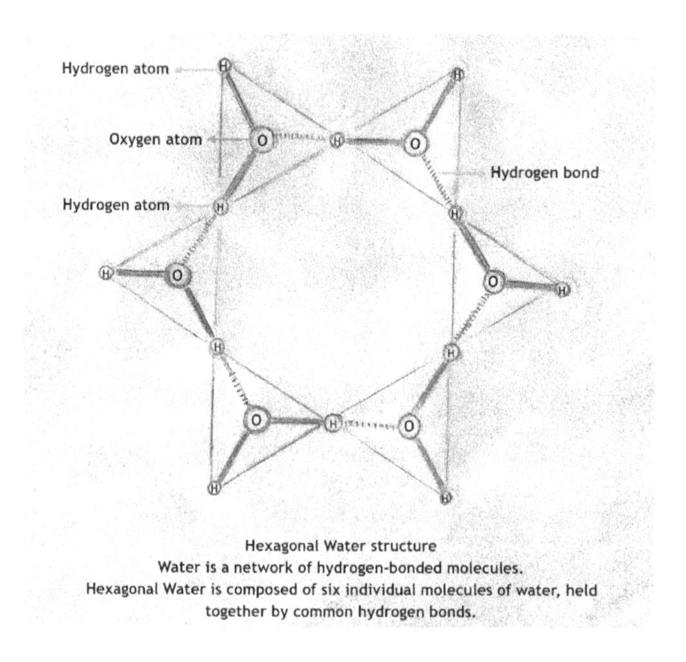


Hexagonal Water structure
Water is a network of hydrogen-bonded molecules.
Hexagonal Water is composed of six individual molecules of water, held together by common hydrogen bonds.

The scientific consensus of researchers familiar with cellular water structure and Nuclear Magnetic Resonance (NMR) is that the water environment surrounding unhealthy cells is less structured and thus able to move more freely than the water environment around healthy cells. According to Dr. Jhon, *hexagonal water forms a protective layer immediately around healthy proteins*. This same type of protection does not exist around unhealthy proteins in the body where pentagonal (five-sided) water clusters are found. Water surrounding normal DNA was found to be highly hexagonally structured, which acts to stabilize the helical structure of the DNA, and forms a layer of protection against outside influences that could cause malfunctions or distortions within the DNA. Also, within the core of each DNA double helix is a column of organized water clusters, which allows it to freely pass through cell walls delivering oxygen, nutrients, protein chains and enzymes; it also removes accumulated toxic materials in a way that unstructured, loosely bonded water (i.e., tap, rain, or mineral) cannot. Science has demonstrated that approximately *90 percent of DNA function involves electromagnetic reception and transmission. This optimal function, underlying every aspect of health, delaying aging, and slowing down virtually every disease process, almost entirely depends on the liquid crystalline core of DNA. Within each living cell, each molecule of protein is surrounded by thousands of molecules of water.* DNA is held together by the hydrogen bonds between the nitrogenous base pairs, which are the same hydrogen bonds that hold together the water around it (water actually stabilizes the entire helical structure by forming hydrogen bonds with the phosphate groups). This means that *DNA cannot be looked at as a freestanding molecule; it is an integral part of a huge organized water cluster.* The communications between these clusters is what we call life. *No life processes can take place without water. The structuring of cellular water is critical to the healthy functioning of every cell.* It is important to understand that it is the microenvironment surrounding the water molecules that influences their arrangement. Thus, cells surrounded by less-structured water may be weaker and more prone to disease and genetic change.

## The Liquid Crystalline Form of Water

In the same way that each of the earth's minerals has a crystalline form (i.e., diamonds are crystalline carbon; emeralds are crystalline beryllium; and rubies are crystalline corundum), *water also has a liquid crystalline form.* The difference between each of earth's minerals and their crystalline form is the way their atoms or molecules are organized. Crystals are pure substances whose atoms or molecules are arranged in an orderly pattern and extend into all three spatial dimensions (height, width, and depth). Not all crystals, however, need to be solid in formation, and an example of this is *liquid crystalline water.* Furthermore, crystal symmetry requires that the atoms or molecules or ions that make up the crystal stack perfectly together with no gaps. Even though the molecules remain mobile in liquid crystalline water, they move together, like a school of fish swimming in the sea. This natural arrangement of water molecules is referred to as structured water, organized water, hexagonal water, or liquid crystalline water, and it forms a repeating, geometric, molecular pattern. This pattern, like all crystal symmetry, adheres to the principle that the atoms or molecules or ions that make up the crystal, stack perfectly together with no gaps. See the diagram below.

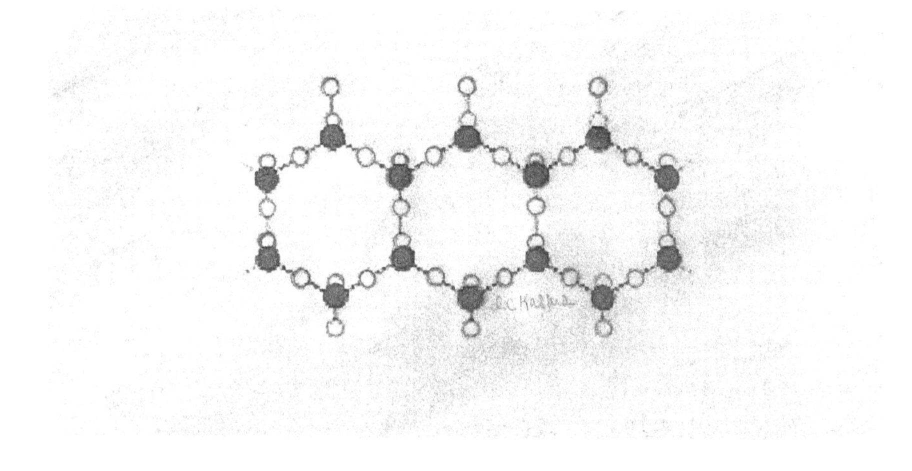

Most of us know that water can be found in either a solid phase (ice), a liquid phase (water), or a gas phase (vapor). As a solid, water molecules arrange into crystalline patterns that we can see when water is in a solid (ice or snow) phase. Yet, research reveals that water also has a liquid crystalline form, which is in between a liquid and a solid, and is made up of organized water molecules. Like solid crystals, organized liquid crystalline structures of water molecules can transmit signals; the repeating pattern provides an efficient pathway for the smooth flow of energetic information. Liquid crystals are flexible and many times more responsive than solid crystals. According to Dr. Gerald Pollack, a professor of bioengineering at the University of Washington and author of several books including *The Fourth Phase of Water*, there is significant evidence for waters liquid crystalline structure. His research has demonstrated water's capacity to form large zones of structured water that is characterized by molecular stability, a negative electrical charge, molecular alignment, an enhanced ability to absorb certain spectra of light, a higher specific heat, and the ability to store energy. When water molecules join to form hexamers, the bond angle widens to 109.5 degrees. This creates greater structural stability and the potential to hold more energy. *What is also new here is the concept that water gets energy from the sun. In other words, the electromagnetic energy of the sun builds up potential energy in the water. Just as plants absorb radiant energy and use it to do work, so too do the water molecules that we drink and the water that comprises our skin and body.* Dr. Pollack's research shows that much of the *water in a healthy human body is in a liquid crystalline/structured state.* Many components of the body are also considered to be liquid crystals, including collagen and cell membranes. These tissues work cooperatively with structured water to create an *informational network involved in cellular communication* that reaches every cell. The capacity of water to store and to transmit information is directly proportional to its structural patterns (characterized by increased hydrogen bonding), and to its coherence (characterized by the degree to which the water can maintain its structure). *The liquid crystalline organization of the human body accounts for the instantaneous transfer of signals and other biological information.* This network also accounts for the photo-perception of our skin, which is composed of 70 percent water, 25 percent protein (mostly collagen), and 5 percent fat.

Dr. Masuru Emoto, in his well-known book, *The Message From Water*, shows that tightly bonded hexagonal water clusters created by structure-making ions (such as calcium, sodium, and zinc) in the middle of the hexagonal structures support healthier cells. Pollutants and toxins are unable to bond with tightly bonded clusters of water molecules to create cellular problems and accumulate toxins. Loosely bonded, pentagonal, or unstructured water clusters and structure breaking ions (such as chlorine and aluminum), however, characterize unhealthy cells where pollutants and toxins can easily bond with water molecules to create cellular problems and toxic accumulation. Dr. Emoto did extensive research in micro clustering of water molecules using Magnetic Resonance analysis technology.

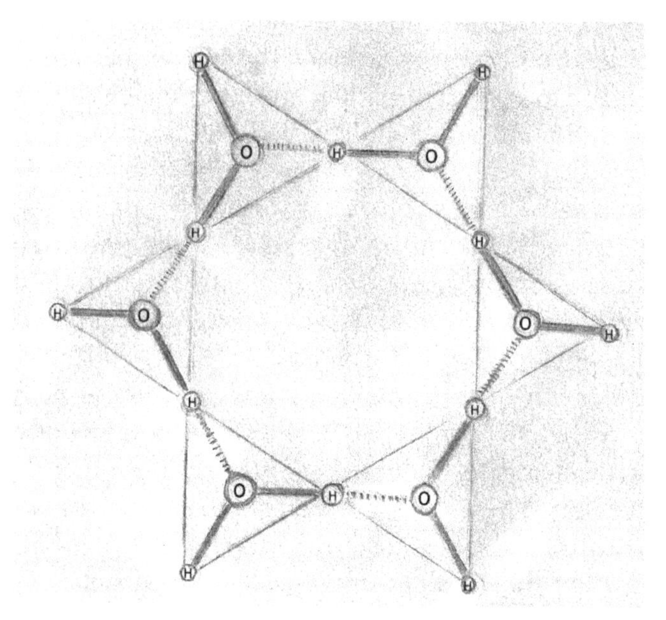


A structured network of six water molecules

The orderly arrangement of water molecules to form a liquid crystal is not a new idea. The Henniker paper (1949) reviews many older works showing massive near-surface (i.e., when water is near the surface of a material) molecular reordering. Furthermore, the idea of precise ordering of water molecules was also advanced by a number of prominent scientists, including Walter Drost-Hansen, James Clegg, and especially Albert Szent-Gyorgyi and Gilbert Ling. Szent-Gyorgyi was a seminal thinker who won the Nobel Prize for discovering vitamin C. A centerpiece of this thinking was the long-range ordering of water, which he regarded as a major pillar in the organization of life. Studies by Dr. Gerald H. Pollack (2013) have shown that organized water molecules emit less infrared energy than disordered water molecules. This is because a crystal's molecular components move around less vigorously and are more stable. It has also been shown that organized liquid water crystals have a higher viscosity than "bulk water."

In summary, structured crystalline water (liquid crystal) hardly resembles unstructured liquid water at all. In addition, the behavior of water depends on its location and its microenvironment; this is clearly seen in how an MRI, which basically measures the properties of the body's water, relies on water's capacity to organize itself differently next to different surfaces. MRI's provide us with a detailed image of structures based on the relaxation properties of protons, which primarily come from water.

Nuclear Magnetic Resonance, or NMR, has the ability to measure molecular size and has been recently used to determine the structure of the water inside the body. This is one of the few ways to verify the size of clusters of water molecules. NMR relaxation time is the time it takes for a molecule to return to its original position after magnetic alignment. Larger water clusters take longer than smaller water clusters to return to their "relaxed" state. NMR measures this time in what is called a line width and the wider the line width, the larger the molecule. Normal tap water has a line-width between 100 and 150 Hz indicating an unorganized state of water and a cluster size between twelve and thirteen water molecules. On the other hand, organized hexagonal water has a line-width measuring between 60 and 70 Hz, which is approximately half the size of unorganized water clusters. The important point here is that the smaller the line width, *the smaller the water cluster* and *the greater its chance of penetrating the hexagonal channels in the cellular membrane to become part of the intracellular water.* The Lawrence Berkeley National Laboratory is one of few laboratories that have been able to physically view (via scanning tunneling microscopy) the existence of water hexamers.

In the *membrane* that surrounds of each of our cells are six-sided, *hexagonal channels or pores,* which form a specialized *intercellular connection that can directly connect the cytoplasm between two cells and allow molecules to pass through.* See the illustration below.

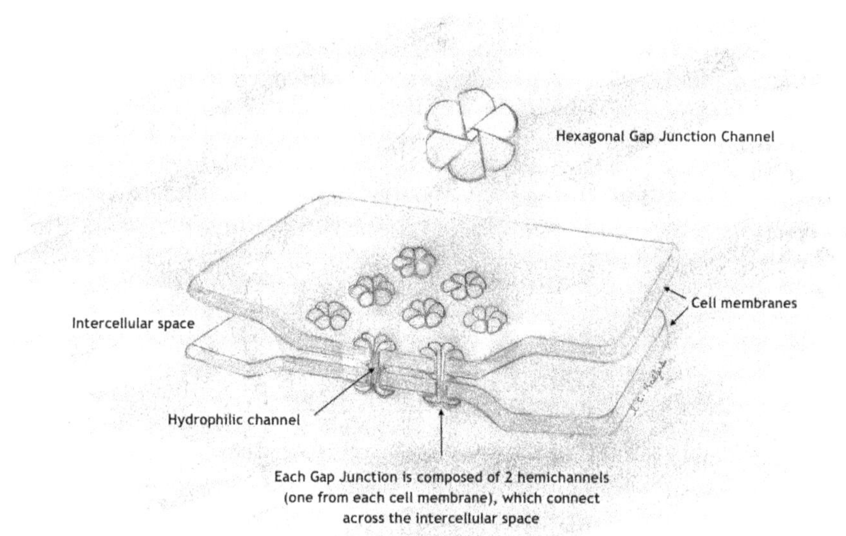


X-ray diffraction shows hexagonal array's in gap junctions between cells

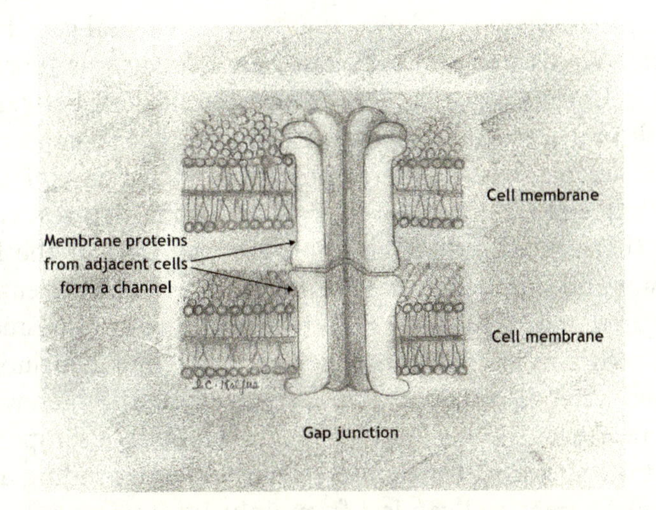


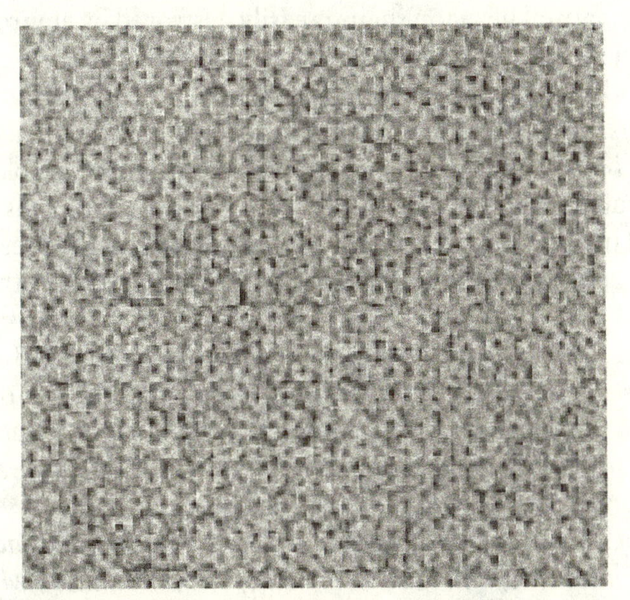

Photo courtesy of US National Library of medicine PMID: 15940850

Electron micrograph of hexagonal gap junctions with central hole diameter (darkly stained center of each hexagon) of about 20 A (one ten-millionth of a millimeter).

These channels are called "Gap Junctions" because Transmission Electron Microscopy has shown that there are actual gaps between cells. Interactions between cells can be stable such as those made through cell junctions. The loss of communication between cells can result in disease, uncontrollable cell growth, and cancer. These junctions are involved in the communication and organization of cells within a particular tissue. Dr. Jhons' published test results[160] verified that structured water penetrates the cells of the human body faster and supplied nutrients and oxygen more efficiently than unstructured water. This also helps with nutrient absorption and waste removal. With over forty years of research, Dr. Jhon concluded that structured water also supports the immune system, which we know communicates with every other system in our body. It's important to note that the stability of structured water varies depending upon its microenvironment and can last from only a few minutes to several days. The natural hexagon shape is easily destroyed by environmental contaminants and modern water treatment processes.

In addition to cells passing their nutrients and information to other cells via gap junctions, cells also acquire their nutrients directly from their surroundings. Larger molecules and ions (molecules or atoms that have an electric charge) pass in and out of cells through their cell membrane; these exchanges require either the presence of a pressure gradient or the cell must expend energy. On the other hand, water and a few other small, uncharged molecules such as oxygen and carbon dioxide easily permeate the cell membrane and diffuse freely (no pressure gradient needed or energy expended) in and out of the cell. *Structured (hexagonal) water has been shown to penetrate cells faster to provide more nutrients and oxygen and to remove cellular waste more efficiently, than unstructured water.*

*Water's unique ability to form and transform, penetrate and dissolve anything it touches, and to collect and deposit information wherever it flows is what helps to give all living things their vital dynamism.* Water molecules readily bond with other molecules as they eagerly seek to mingle with other elements and pick up good things such as oxygen and nutritious minerals. However, they also pick up bad things like toxic pollutants, poisonous chemicals, and disease-causing organisms

---

[160] Jhon MS, Pagman MJ, The Water Puzzle 2004.

that can enter into our bloodstream. This is why healthy water, which is saturated with oxygen and minerals, is so vital. Ironically, many of us like to drink soda, which contains dissolved carbon dioxide gas under pressure. We breathe in oxygen and exhale carbon dioxide, which is one of the waste products of the human body. Drinking carbonated beverages add large quantities of the very chemical the body is trying to get rid of. This carbon dioxide from soda will push some of the oxygen out of the cell resulting in a temporary hypoxia (lack of oxygen), which induces faulty metabolism. This eventually results in diseased organs.

In nature, water becomes oxygenated when flowing over stones, various objects, and waterfalls. Some cultures call this "living water" because it is compatible with life (i.e., freshwater fish can live in it). On the other hand, distilled water and carbonated water are called "dead waters" because neither fish nor plants can live in them. In addition to oxygen enrichment, natural water also becomes enhanced by light from the sun and electromagnetic stimulants from granite type rocks and the earth's natural magnetic field. The end result is a hexagonal structure carrying oxygen, energy, and selected ionic minerals deep into the cellular environment within seconds. Some wavelengths of light penetrate deeply into our bodies. For example, visible red light (630–700 nm) penetrates tissue to a depth of about 10 mm, which is close to the skin's surface, while infrared light (800–1,000 nm) penetrates to a depth of about 40 mm. Interestingly, we can easily see how white light from a flashlight, when placed on the back of your hand along the fingers, will easily penetrate all the way to the other side.

The diverse tissue and cell types in the body all have their own unique light absorption characteristics; that is, they will only absorb light at specific frequencies (wavelengths) and not at others. This is why, for example, skin layers, which are high in water content and blood flow, absorb visible red light readily, while calcium and phosphorous (involved in muscle movement and skeletal composition) absorb light of a different wavelength such as infrared light. As previously mentioned, our skin is capable of discriminating between certain electromagnetic messages (light is part of the electromagnetic spectrum), a phenomenon referred to as cutaneous (skin) photoreception. The external ear, in particular, will only absorb particular frequencies of light when the body is healthy and will absorb additional frequencies only when there is a corresponding ailment in specific body tissues. Pulsing light at specific rates (known as the Nogier frequencies) can dynamically reinforce the therapeutic effects of light therapy and dynamically reinforce the healthy function of body tissues.

Minerals such as calcium, sodium, zinc, iron, and copper aid in the formation of hexagonal water clusters, while potassium, chloride, fluoride, aluminum, magnesium, and sulfide destroy the hexagonal clustering of water molecules. Water, and especially healthy water is, therefore, vital to our health and well-being. Dr. Yunjo Chung M.D., and author of *The Cleansing Side of Medicine* discussed how research has shown that *the main cause of aging is the accumulation of waste products in the body and only proper amounts of water can clean your body's cells.* In his opinion, drinking plenty of water (but "everything in moderation") along with moderate exercise prevents the buildup of toxic waste in your body, thus expelling the cause of many diseases. Researchers are aware that many diseases are associated with an unusual build-up of waste chemicals around cells, such as in Alzheimer's disease where it is believed that aluminum (which is nonessential to life) is the main culprit. In particular, it is the aluminum in drinking water that is of primary concern.[161] There are literally hundreds of scientific studies that show a relationship between the metal aluminum and its neurotoxicity to living systems. Thus, the quality of the water that we drink along with maintaining proper hydration and circulation are all key to maintaining health. I was taught this in the 1980s while studying biochemistry, and I have often used the analogy of a fast funning stream versus a pond or a lake. Water that is moving quickly does not accumulate much

---

[161] Martyn CN, Osmond C, Edwardson JA, Barker DJP, Harris EC, Lacey RF. Geographical relation between Alzheimer's disease and aluminium in drinking water. *The Lancet.* 1989;1(8629):59–62; Flaten TP. Aluminium as a risk factor in Alzheimer's disease, with emphasis on drinking water. *Brain Research Bulletin.* 2001;55(2):187–196; Rondeau V, Jacqmin-Gadda H, Commenges D, Helmer C, Dartigues JF. Aluminum and silica in drinking water and the risk of Alzheimer's disease or cognitive decline: findings from 15-year follow-up of the PAQUID cohort. *American Journal of Epidemiology.* 2009;169(4):489–496.

debris, while water that is still, does. According to published data by Dr. Batmanghelidj, M.D., the author of several books on chronic ailments, *dehydration is the underlying cause of many chronic diseases.*

As previously mentioned, to understand the language of healing, we must first realize that every form of life we know of depends upon frequencies, which are often measured by the number of times an object vibrates, or moves back and forth, in one second. Each water molecule has three fundamental vibrations where the oxygen and hydrogen molecules stretch and bend. For nonlinear molecules (such as water), the general rule is "3*N – 6". "N" stands for the number of atoms in a molecule. For water, we have three atoms (2 hydrogen and 1 oxygen atom). Thus $(3 \times 3) - 6 = 3$; this means that a molecule of water will exhibit three functional vibrations or frequencies. The frequency of the vibrations depends upon temperature. It is light frequency, in particular infrared light that provides the energy needed to organize the water molecules in living cells into a liquid crystalline pattern that is central to both the informational network involved in cellular communication as well as to form a protective layer immediately around healthy proteins. *The organization of water molecules is, therefore, central to our health.* It is also interesting to see that the number "three" often plays a key role in auricular therapy; each water molecule has *three* fundamental vibrations and water constitutes up to 75 percent of our body weight as a newborn, averages over 50 percent in adults and constitutes 99 percent of our molecules; every cell in our body is derived from one of *three* embryonic layers of cells; the ear is divided into *three* regions that correspond to each of these cell layers; it takes *three* days for a neurovascular point in the ear to renew itself after it has been "destroyed" by an acupuncture needle; and Dr. Paul Nogier discovered *three* phases on the ear. The phases are transient neurological representations of the body on the ear depending upon whether a condition is acute, degenerative, or chronic. In other words, each area on the ear that corresponds to an area of the body will have *three* different locations depending upon the condition of the ailment.

# Chapter 26

## How Ear Acupuncture Relieves Pain

To summarize pain transmission and how auricular therapy affects pain, we only need to understand and acknowledge that acupuncture stimulates nerve fibers that send impulses to the spinal cord and the brain (as well as the midbrain, hypothalamus, and anterior/posterior pituitary) to cause analgesia. Enkephalins and dynorphins block pain transmission at the spinal cord level. The presynaptic inhibition probably works by reducing calcium current inflow during the action potentials resulting in reduced release of the pain transmitter. The midbrain uses enkephalins to activate the raphae descending system, which inhibits spinal cord transmission by a synergistic effect of the monoamines, serotonin, and epinephrine. The midbrain also has a circuit that bypasses the endorphinergic links at high frequency stimulation. Finally, at the hypothalamic/pituitary center (inferior concha of the auricle), the pituitary releases endorphins into the blood and cerebrospinal fluid to cause analgesia at a distance. The hypothalamus/pituitary is activated at low frequencies. What is clinically significant is that when either an acupuncture needle or a bipolar electrode is placed in a tender area or at a precise point of electrical disparity at the site of injury, you are maximizing the segmental (spinal cord) stimulation and local analgesia. Stimulation that is remote from the painful site will activate the midbrain and hypothalamus/pituitary producing analgesia throughout the body. The auricle provides access to both segmental and extrasegmental analgesia. Han, J.S., 2001, demonstrated that different endorphins are activated by different frequencies of electroacupuncture. Han concluded that 2 Hz activates enkephalin synapses, whereas 100

Hz electroacupuncture activates dynorphin synapses. The opiate peptides enkephalin and dynorphin are two subfractions of the larger polypeptide molecule known as beta-endorphin. Note that low frequency <10 Hz, high intensity uA works through the endorphin system for slower onset and longer duration of pain relief with cumulative effects while high frequency 10 –160 Hz bypasses the endorphin system for fast onset and short duration without cumulative effects.

Low frequency, <10 Hz, can be given daily or two to three times a week for long-term cumulative effects. Lower frequencies (<10 Hz) most affect enkephalins, endorphins, and visceral and somatic disorders, whereas higher frequencies (10—160 Hz) affect dynorphins and neurological dysfunctions. Often times, western providers treat infrequently (one to two times per week) and complete treatment after five to ten sessions. In Asia, patients may be treated daily for a month, then weekly for six months, and the results reported are excellent.

Being so densely populated with over two hundred auricular acupuncture points on each auricle, one must question if the points are really there or are they contrived and inaccurate. The points are true and specialists from all over the globe have mapped them out. *However, it is sometimes difficult to evaluate sham or fake points because our skin is so densely covered with points (some of which are yet to be discovered) that many areas will elicit measurable changes in our body even though they have yet to be scientifically proven.* The first English language texts of the Chinese system of ear acupuncture included the works of Helean Huang (1974), Mario Wexu (1975), and Nahemkis and Smith (1975). In the 1980s and 1990s, several European countries including France, Germany, and Italy, translated the Chinese texts. The international expert Li-Chun Huang also wrote a textbook on auricular diagnosis and ear acupuncture treatment in 1996. In 1957, the French expert and pioneer of auriculotherapy, Dr. Paul Nogier, was the first to map out the ear in an organized pattern that corresponds to the arrangement of the actual body (versus the preciously scattered array of different points). In 1968, Dr. Nogier, along with Dr. Rene Bourdiol, introduced the world to the embryological correspondences of the ear in the *Handbook to Auriculotherapy.* And in the United States, Dr. Terry Oleson published the Auriculotherapy Manual in

1990. The Auriculotherapy Certification Institute was established in 1999 to certify practitioners who demonstrate a high level of mastery in the field. As you can see, therapeutic ear points have been well documented by world-renowned experts for decades.

In research, we want to determine if something is authentic or fake, a sham, not what it is purported to be. Numerous studies on acute pain show that acupuncture analgesia from needling true points is far superior to acupuncture analgesia from needling sham (fake) points. For chronic pain, true points are effective in 85 percent of cases.[162]

There are studies that have found little or no significant differences in the outcomes between sham (or fake) acupuncture and true acupuncture. However, there are many more electrically active points than what were previously accepted by the World Health Organization (WHO) or by national acupuncture organizations. It is estimated that, in China, for example, more than 2,000 new electrically active points have been discovered and defined. This means that, when an observer uses a supposedly sham (fake) treatment point, believing that he or she is treating an area that is presumably inactive, the researcher does not actually know if it is a nonregistered (not yet officially identified) point or an actual point sometimes referred to as a satellite point, which can also have some clinical efficacy.

In auricular medicine, *bipolar detection is the only scientifically proven way to locate an electrically active point on the ear.* And for those of us who use Nogier's three-phase approach, this means that the treatment points move depending upon whether the patient's condition is acute, degenerative, or chronic, which means there are hundreds of possible sites on the ear alone that may or may not be active and any one time. *Using FDA approved bipolar electrode detection technology, along with checking for area's that are sensitive to the touch, there is no doubt as to the presence or absence of an ear point that may need treatment.*

---

[162] Vincent CA, Richardson PH, 1989; Vincent CA, Lewith G, 1995.

# Chapter 27

# Beauty Is More Than Skin Deep (The histology of electrically active ear points)

Do the electrically active/neurovascular points on the ear have a unique anatomic structure? Yes! From the work of Senelar, Auziech, and Terral (1970–1980) that looked at ear points with lower cutaneous electrical resistance (CER) in humans and examined them under optical microscopy, they found specific histologic features under the points of lower cutaneous electrical resistance to include an arteriole, a venule, a lymphatic vessel, and a free nerve ending. Myelin-sheathed nerve fibers are distributed among the vascular elements and may come into close proximity to the vascular structures. This coexistence of nerves and thin-walled vessels in close proximity is NOT a random structuring, and we should note the release of hormonal or related factors by endocrine constituents following stimulation of the points. The neurovascular points or ear points are detected by the use of instruments that measure variations in electrical resistance (bipolar electrodes ensure the highest level of precision as they simultaneously measure both electrical conductivity and resistance differentials within a prescribed radius), and they are detectable when there are peripheral functional disorders. These formations are called neurovascular complexes and have a consistent pattern below the points used in auricular therapy. These neurovascular complexes are related to homeostasis, maintenance of the skin, and control of the blood supply to internal organs, playing a specific role in the thermoregulation process. A report by Heine[163] revealed that 80 percent of electrically active ear points correlate with perforations in the superficial fascia that may correlate to meridians or biologic energy channels. Nerves are very important to electrically active ear points because it's been demonstrated that you can abolish neurovascular point by injecting local anesthetic into the point before stimulation begins.[164] Research has clearly identified as many as three hundred acupoints either situated on or very close to nerves, while an almost equal number are on or very close to major blood vessels that are surrounded by small nerve bundles and peripheral nerves.

163 Heine H, 1988.

164 Chiang CY, Chang CT et al, 1973; Pomeranz B, Paley D, 1979.

# Chapter 28

## Physiological Features Unique to Auricular Reflex Points

What are the unique physiological features of electrically active ear points? The external ear is unique and unlike any other area on your body. Only in the ear does a point with a Mega Ohm Differential correspond to tissue pathology or imbalance. In other words, when a point on the ear has a different electrical resistance/conductance relative to the area immediately surrounding it, this is a clear signal that there is a problem in the corresponding part of your body. When there is no "signal" or detectable differentiation in electrical resistance/conductivity at a point relative to its immediate surroundings, this indicates health and balance in the corresponding part of the body. Using the Stim Flex 400A, the gold standard in sophistication and precision, to detect active points, the center electrode is 0.67 mm diameter and the outer surrounding electrode ring is 3.0 mm diameter providing pinpoint location of a Mega Ohm differential point within the scanned area. Unlike the rest of your body that normally has a DC resistance of 2 million ohms, your body acupuncture points have a DC resistance down to 50,000 ohms. Several studies have demonstrated that active acupoints on the ear have a mega ohm differential relative to their immediate surroundings.[165] The first double-blind assessment to validate scientifically the somatotopic pattern of electrically active points on the auricle was conducted by

[165] Nakatani Y, Yamashito K, 1977; Nogier PFM, 1972; Oleson TD, Kroenig RJ, Bresler DE, 1980.

Dr. Terry Oleson.[166] Dr. Oleson compared diagnoses made with a point finder (which detects differences in electrodermal conductivity and electrodermal resistance) on the ear with diagnoses made on the same patients by means of a western medical work up. The researches were "blind" to the western diagnosis to ensure that no clues were available to them. Auricular diagnosis was determined by numerically rated levels of tenderness to a palpating probe and by the quantified electrical conductivity of the skin. Specific areas of the auricle, which corresponded to different musculoskeletal regions of the body, were examined. The correlation between ear diagnosis and the established medical diagnosis was a statistically significant 75.2 percent.[167] When the pain was located on only one side of the body, electrical conductivity was significantly greater at the auricular point on the ipsilateral ear than at the corresponding area of the contralateral ear. These results support the concept that specific areas of your ear are related to specific areas of your body. Thus, *there is a clear correlation between the bioelectric properties of the external ear and the presence of disease or dysfunction of the corresponding area of your body.* It is also clear that most active auricular points have a lower electrodermal resistance than the surrounding skin, and this goes together with an increase in biophoton emissions. Sweating has a profound effect on skin resistance. Stress activates the sympathetic nervous system, which causes sweating and a drop in skin resistance. Although sweating occurs equally at acupuncture points and over the surrounding skin throughout our body, *tissue samples of the skin overlying the ear, paradoxically, do not reveal the presence of any sweat glands; this indicates that some other process must account for the spatial differences in skin resistance to the flow of electricity at electrically active ear points.* Research by Jaffe[168] demonstrated that your skin has a resting potential across its epidermal layer of 20–90 mV (outside negative, inside positive) and active ear points, having low resistance, may short-circuit this "battery" across your skin and give rise to a source of current. In other words, electrically active ear points often provide a path of least resistance for currents driven by

---

[166] Oleson T, Kroening R, Bresler D, 1980.

[167] Oleson et al., 1980.

[168] Jaffe L, Barker AT et al, 1982.

the 20–90 mV resting potential which exists across your entire skin. Jaffe et al.[169] also showed that a cut in the skin produces a current of injury, which is due to short-circuiting of the skin battery. Insertion of acupuncture needles into the skin produces a decrease in local skin resistance that lasts for one to two days and produces a current of injury that is a benefit to tissue growth and nerve regeneration.[170] In 1961, the French Physician, Niboyet, wrote that acupuncture points have a much lower electrical resistance than the surrounding skin.[171] Normally dry skin has a resistance in the order of 200,000 to two million ohms. At acupuncture points, this resistance is down to 50,000 ohms.[172] In 1977, it was discovered that the distribution of these points with lower resistance was exactly the same as the localization of the Chinese acupuncture points discovered thousands of years ago.[173] Thus, *the importance of precisely locating mega-ohm differential points in auricular therapy and providing the proper wave (negative square wave vs. positive square wave), frequency, current, and duration of treatment cannot be overstated.* Although most active auricular points are electroductive, it is equally important to precisely locate the electodermally resistant points. The key point here is the differential between the point and the adjacent skin and to treat accordingly. Your active ear points have both target specific effects and total body effects.

I am often asked about meridians or pathways through which the life energy known as qi (pronounced "chee") flows according to Traditional Chinese Medicine. The traditional concepts of the meridian system, the foundation of traditional acupuncture theory, have been studied from the perspective of morphogenesis. The relationship between the meridian system and embryogenesis has been noted for decades.[174] The development of organizing centers, which are high electric conductance points on the body surface[175]

---

169 Ibid.

170 Borgens RB, Variable JW, Jaffe LF, 1979; Wu C, 1984.

171 Niboyet, 1961.

172 Becker, 1976.

173 Hyvarimen, 1977.

174 Mann F, 1971

175 Shang C, 1989

in the growth control system, precedes the development of the nervous system and other physiological systems. The formation and maintenance of all other systems is dependent on this electrically active system. The meridian system is thought to originate from this network of organizing centers. It's like our genetic blue print. Consequently, it overlaps and interacts with the other systems but is not simply part of them. Incredibly, auricular acupoints offer access to this physiological system that preceded and interacts with all other systems. Active neurovascular ear points, which are focal points of detectable mega-ohm differentiation, current density, and a high concentration of gap junctions (for intercellular communication), are not limited to the obvious and readily accessible extreme points of surface curvature (organizing centers are expected to be at the extreme points of curvature on the body surface, such as the locally most convex or concave points). Rather, they can also be related to internal structures. Thus, the undifferentiated, interconnected cellular network represented as meridians, is not limited to your skin. It is suggested that the existence of this network of cells such as the germ cell[176] comprises part of the meridian system and provides important regulatory functions.[177] It is likely that there is a hierarchy in the degree of cell differentiation and function in the meridian system. The more superficial meridians and acupuncture points are more differentiated while the less differentiated remain well below the surface and constitute the central core of the regulatory meridian system.

As previously mentioned, our epithelia (skin) usually maintains a 30–100 mV-voltage difference.[178] This voltage is the potential difference across cell layers, not membrane potential. An active auricular acupoint, one that shows a distinct differential in electrical conductivity relative to the surrounding skin and indicates an underlying problem that corresponds to the area of the body that the point relates to, will also have a converging point of surface current. This is a singular point of abrupt change in electrical current flow. A

---

176 Azizhan RG, Caty MG, 1996; Kountakis SE, Minotti AM, Maillard A, Stiernberg CM, 1994; Dehner LP, 1990; Kretschmar CS, 1997

177 Nichols CR, Timmerman R, Foster RS, Roth BJ, Einhorn LH, 1997

178 Jaffe LF, 1977

singular point is a point of discontinuity as defined in mathematics and indicates a point of abrupt transition from one state to another. Small disruptions around singular points can have decisive effects on a system. Thus, stimulating auricular points, which are singular points in the skin's surface bioelectric field, can elicit impressive changes.

During the earliest stages of our development, the fate of a larger region is often controlled by a small group of cells known as organizing centers.[179] Organizing centers are the high electric conductance points found on the surface of our skin that we call acupuncture points. On the external ear, these points are only active (exhibit a differential in electrodermal conductivity/electrodermal resistance relative to the skin around it) when there is an underlying problem with the area of the body that the acupoint, or organizing center, corresponds to. The high electrical conductance phenomenon is further supported by the finding of high density of gap junctions (specialized intercellular connections between cells; e.g., synapses) at the sites of organizing centers.[180] Organizing centers are singular points in the morphogen gradient and electromagnetic field.[181] Morphogen gradients are what generate cell differentiation. Most morphogens are secreted proteins that *signal between cells.* And cell signaling is part of a complex system of communication that governs basic cellular activities and coordinates cell actions. This is the basis of development, tissue repair, immunity as well as normal tissue homeostasis. As noted above, biophotons also affect the electrical activity of cells via signaling processes using speed of light transmissions through the air rather than slower communications that occur by either physical contact between cells, through fluid (hormones/chemicals), or interstitial (electrical synapse as with neurons). Both acupuncture points and organizing centers have high electric conductance, current density, a high density of gap junctions, and can be activated by nonspecific stimuli. A therapeutic effect of acupuncture can be achieved by a

[179] Meinhardt h, 1982

[180] Laird DW, Yancey SB, Bugga L, Revel Jp, 1992; Yancey SB, Biswal S, Revel JP, 1992; Coelho CN, Kosher RA, 1991; Meyer RA, Cohen MF, Recalde S, Zakany J, Bell SM, Scott WJ fr, Lo CW, 1997

[181] Shang C, 1989

variety of stimuli[182], including light, electrical stimulation, needling, temperature variation, laser, and pressure. Similarly, *almost all the extreme points of the body surface curvature are acupuncture points.* These similarities between acupuncture points and organizing centers indicate that acupuncture points originate from organizing centers. And the external ear is unique in that each of the three embryonic tissue layers from which every cell in our body is formed, is neatly organized and represented, giving us access to a complex system of cell-to-cell communication.

Why is it that your ear is so densely populated with acupuncture points and has over 10 percent of all body points? The distribution of acupuncture points and organizing centers (small groups of cells that have high electric conductance) is related to the morphology or shape of an area on your body; and the ear has the most complex surface morphology. The specific locations of over two hundred auricular acupuncture points on each ear have been well documented with over twenty-five books devoted to auricular points alone. According to the World Health Organization, forty-three auricular points have proven therapeutic value.[183] Although the outer ear has been thought to have no significant physiological function other than gathering sound waves and directing them to the auditory canal and eardrum, auricular morphology is one of the most sensitive indicators of injury and damage in other organs. It is recommended in a standard textbook of pediatrics that any auricular anomaly should initiate a search for malformations in other parts of the body.[184] Auricular specialists have emphasized the diagnostic value of visually examining the external ear.[185] The *New England Journal of Medicine* published studies[186] that correlated diagonal ear lobe creases with coronary problems. *The presence of a crease running diagonally from the intertragic notch to the bottom of the lobe was more predictive of the occurrence of a coronary problem than knowing either the patient's blood pressure level or serum cholesterol level.* And there is a direct correspondence

---

182 Vickers AJ, 1996; Altman s, 1992

183 World Health Organization, scientific group, 1991

184 Cotton RT, 1996, Rudolph's Pediatrics

185 Kvirchishvili, 1974; Romoli and Vettoni, 1982; Huang. 1996

186 Lichstein et al., 1974; Mehta and Homby, 1974

between the depth of the crease on the ear lobe and the severity of the health problem. Likewise, the occurrence of discolorations and/or dandruff like areas on the ear suggests a high probability of some type of acute or chronic pathology in the correspondent part of the body. These spots will gradually go away when the health of the related body area improves.

Since the 1950s, it has been discovered and confirmed by researchers in several countries with refined techniques[187] that most acupuncture points correspond to the high electric conductance points on the surface of the skin.[188] Quantified examinations of the electrical properties of the skin have provided the most objective demonstration of scientific validity of acupuncture points[189]. Reichmanis, Becker, and Marino[190] systematically verified that the electrodermal resistance at acupuncture points is significantly lower than in surrounding tissue. Kawakita and associates provided objective support for the organocutaneous reflexes described by Dale[191], wherein low resistance auricular acupuncture points appear in response to pathology of the corresponding pathology. Yet, merely looking at electric conductance points as a focal point to detect and treat an underlying condition is misleading. The key feature here is the differential between the electrical conductivity of the point to be treated and the electrical resistance in the area immediately surrounding the point, or vice versa. If a highly conductive point is detected without knowing the conductivity/resistance of the surrounding skin, the net electrical differential may be zero, thus negating the effect of the treatment and giving a false positive (i.e., it was not an active point). The only scientifically proven way to detect and therefore treat an active point is to use a bipolar electrode, one that measures the differential between the acupoint and the

---

187 Pomeranz B, 1997

188 Comunetti A, Lassage S, Schiessl N, Kistler A, 1995; Bergsman O, Wooley-Hart A, 1973; Wensel LO, 1980; Nakatani Y, Yamashita K, 1977; Reichmanis M, 1988

189 Becker R, Reichmanis M, Marino A, Spadaro J, 1976; Bergsmann O, Hart A', 1973; Oleson T, Kroening R, Bresler D, 1980; Reichmanis M, Marino A, Becker R, 1975

190 Reichmanis M, Marino A, Becker R, 1975

191 Dale R, 1993

skin surrounding it. Simply put, the most effective points have the greatest deflection/mega-ohm differential. If the center electrode is positive (conductive) and the outer electrode is negative (resistant), an electrical differential will be present indicating an active point that can be treated by delivering a positive square wave from the outer probe to the center probe. Likewise, if the center electrode is negative (resistant) and the outer electrode is positive (conductive), this also indicates an active point that can be treated by delivering a negative square wave from the outer electrode to the center electrode. No matter what the electrical conductance or resistance is on the probes touching the skin, if the bipolar electrodes are equally resistant or conductive, the acupoint is not considered active. Famed neurologist Dr. Paul Nogier[192] noted that bipolar detection of ear reflex points not only led to the discovery of low electrical resistance points but also allowed for the *identification of high skin resistance ear points* that can be *as significant* in the selection of ear acupoints to be used for successful treatment of health conditions as low electrical resistance points. Simultaneously, both Dr. Jim Shores and Dr. Terry Oleson, world-renowned experts in auricular therapy, have also recommended its use since the 1970s.

Auricular acupuncture points not only harbor high electrical conductance and resistance points, they also have a high density of gap junctions, which are hexagonal protein complexes that form channels between adjacent cells. A gap junction is an opening from one cell to another and the connection forms a channel that acts like a tunnel for the movement of molecules from one cell to the other. They facilitate intercellular communication and increase electric conductivity. Acupoints have also been found to have higher temperature,[193] metabolic rates, and carbon dioxide release.[194]

At the early stages of embryogenesis, gap junction mediated cell-to-cell communication is usually diffusely distributed, which results in the entire embryo becoming linked as a fusion of cells. As development progresses, gap junctions become restricted at discrete boundaries, leading to the subdivision of the embryo into

---

192 Nogier P, 1972. Treatise of Auriculotherapy.

193 Zhang D, Fu W, Wang S, Wei Z, Wang F, 1996.

194 Eory A, 1984.

communications compartment areas.[195] These high conductance boundaries are also major pathways of bioelectric currents and likely to be the precursors of meridians (energy pathways). It was proposed that interconnected cells in the meridian system remain under-differentiated and maintain their regulatory function in a partial embryonic state.[196] Thus, the meridian system likely originates from a network of organizing centers (singular points with a different electrical field than its surroundings), and remember that organizing centers are high electrical conductance points that, when found on the skin become acupuncture points.

As the electric conductance of organizing centers changes with development, the conductance of acupuncture points also changes and correlates with the physiological change[197] and pathogenesis.[198] The fact that the change in the electric field precedes structural change during development[199] and that manipulation of the electric field can affect the change[200] may help with the diagnosis and treatment of many diseases. Stimulation of organizing centers via auricular therapy is likely to be involved in restoring proper form and function to the organizing center network that will restore health to the surrounding tissues. We know that acupuncture can speed up the wound healing process[201] and cause an exaggerated systemic wound healing and stress response.[202] The response can include excessive release of endorphins, which stimulates epithelial cell growth[203] as well as analgesia. Other neurohormonal factors induced by auricular therapy such as serotonin[204] and adrenocorticothyroid hormone[205]

---

195 Lo CW, 1996.

196 Cui H-M, 1988; Shang C, 1989.

197 Comunetti A, Laage S, Schiessel N, Kistler A, 1995.

198 Saku K, Mukaino Y, Ying H, Arakawa K, 1993; Oleson TD, Roenig RJ, Bresler DE, 1980.

199 Nuccitell R, 1988.

200 Altman S, 1992.

201 King GE, Scheetz j, Jacob RF, Martin JW, 1989.

202 Wong WH, Brayton D, 1982; Lin MT, Liu GG, Song JJ, Chen YF, Wu KM, 1980.

203 Kishi H, Mishima HK, Sakamoto I, Yamashita U, 1996.

204 Cheng RS, Pomeranz B, 1979.

205 Malizia E, Andreucci G, Paolucci D, 1979.

also have effects on growth control and healing.[206] In auricular therapy, the stimulation of a singular point (acupuncture point) may not directly provoke a pathological process but may indirectly adjust the process and restore normal function by activating the network of organizing centers in your body. Remarkably, stimulation of acupoints is bidirectional in that it will either suppress hyperfunction or stimulate hypofunction. The activation of organizing centers to elicit a normalizing function is less likely to cause the side effects resulting from directly antagonizing a pathological process that often overlaps with normal and beneficial physiological processes. Therefore, *stimulation with a bipolar electrode at the appropriate frequency and amplitude will place a positive (anode) pulse to inhibit a corresponding function or a negative (cathode) pulse to enhance a function.* This polarity effect is similar to the finding that cell growth is enhanced toward the cathode (negative charge) and reduced toward the anode,[207] which is consistent with the similarities between auricular acupuncture points and their embryological origins as organizing centers. This is real! And if real is what we can feel, smell, taste, and see, then "real" is simply electrical signals interpreted by our brain.

---

[206] Pakala R, Benedict CR, 1998.

[207] Nuccitelli R, 1984; McCaig CD, 1987.

# Chapter 29

# Identifying Neural Correlates with Auricular Therapy Using Brain Imaging

There is direct evidence showing that auricular stimulation specifically effects brain activity as measured by fMRI. Dr. David Alimi of France[208] demonstrated that stimulation of specific areas of the auricle led to selective changes in fMRI responses in the brain. Additional studies[209] also demonstrated fMRI changes with auricular acupuncture. Dr. Z.H. Cho[210] showed that experimentally induced *pain leads to increased fMRI activity* (increased oxygen) in the cingulated gyrus of the limbic brain, in the thalamus, and in the periaqueductal gray (PAG). *Following acupuncture stimulation, each of the areas of the brain that showed elevated fMRI activity on Dr. Cho's subjects was dramatically reduced, suggesting that acupuncture inhibits the neurons in the brain that respond to pain.* These cortical projections of signals following stimulation of acupoints demonstrate that many of the effects of auricular therapy are redirected through the central nervous system (CNS). This concept of CNS involvement is supported by a considerable body of experimental evidence into the area of acupuncture analgesia, where correlates have been observed

---

208 Alimi, 2000; Alimi et al., 2002

209 Alimi D, Geissmann A., and Gardeur D. Med Acupunct. 2007; 13(2): 18–21; Liboni W., Romoli M., Allais G, et al. Acti XXII Congresso Nationale Soc Ital Rigles Ago Auri (SIRAA: November 16–17, 2007).

210 Cho and Wong, 1998; Cho et al., 2001.

between acupoint stimulation analgesia onset, with release of a variety of neurotransmitters, endogenous opioids, and hormones in the brain, spinal cord, and peripheral circulation.[211] See fMRI images below of the brain from a series of studies by Dr. Z. H. Cho.

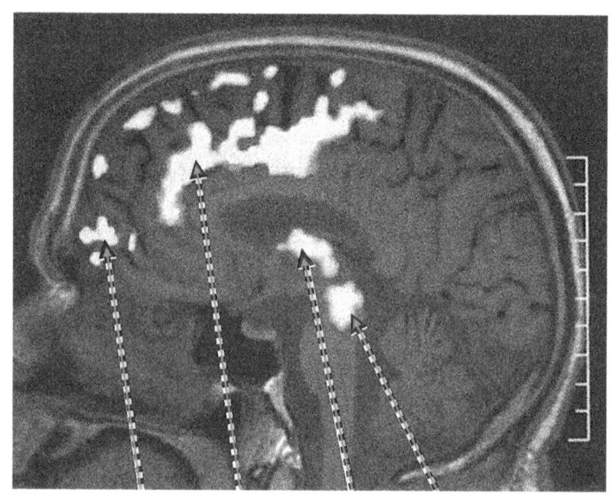

Pre-Frontal Cortex Thalamus

Cingulate Gyrus Periaqueductal gray

fMRI image of brain courtesy of Dr. Z. H. Cho

(Brain activity to pain / **before** - acupuncture)

[211] Filshie J, White A, 1998; Han JS, 1993; Pomeranz B, 1987; Pomeranz B, 1996: Takeshige C, Sato T, Mera T, Hisamitsu T, Fang J, 1992.

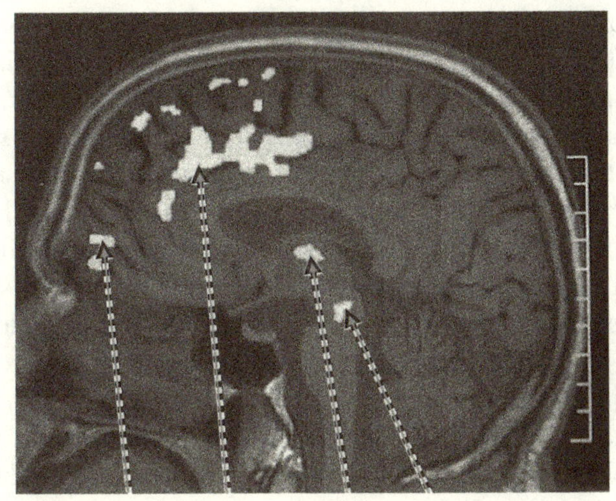

Pre-Frontal Cortex Thalamus

Cingulate gyrus Periaqueductal gray

fMRI image of brain courtesy of Dr. Z. H. Cho

(Brain activity to pain / **after** - acupuncture)

Note: The lighter areas are smaller (less excitation) following acupuncture

Several imaging techniques, including Positron Emission Tomography (PET), Single Photon Emission Computed Tomography (SPECT), and functional Magnetic Resonance Imaging (fMRI) are used to measure neural activity as reflected in the uptake of cerebral blood glucose or increased cerebral blood flow. Unlike most other cells, nerve cells have little stored glucose and rely on the uptake of blood sugar as well as blood oxygen to sustain their high levels of activity. Since neural activity is critically dependent on oxygen derived from cerebral blood supplies, fMRI is an extremely sensitive means of measuring deoxyhemoglobin. Studies in Boston and Taiwan[212] have demonstrated that a wide variety of brainstem, midbrain, and cerebral cortical structures show reproducible patterns of increased or decreased activity related to acupoint stimulation.

212 Chan SH, 1984; Hui KKS, Liu J, Wu MT, Wang K, 1996; Hui KKS, Liu J, Makris N, Gollob RL, Chen AJ, Moore CL, Kennedy DN, Rosen BR, Kwong KK, 2000.

The stimulation of acupoints evokes "signals" that are projected to the CNS that in turn correlates to acupuncture analgesia via the release of neurotransmitters, endogenous opioids, and hormones in the CNS and peripheral circulation. *One of the miracles seen with auricular therapy is how this therapy not only provides pain relief, but goes well beyond to promote healing.* This is due to both the higher cortical involvement as well as the survival (fight or flight) functions. Using SPECT imagery, studies of chronic pain patients showed a marked left/right asymmetry in the pretreatment blood flow of the thalamus, a major site in the neural integration of pain sensation. Following acupuncture treatment, all patients reported pain relief and the thalamus asymmetries were significantly reduced. Note that the control subjects showed no blood flow asymmetries either before or after acupuncture.

## Chapter 30

# Identifying Endorphin and Neurochemical Correlates Following Auricular Stimulation

Endorphins, endogenous morphine, enkephalins, and dynorphins are natural pain relieving neurotransmitters. Pain and stress are the two most common factors leading to the release of endorphins. They interact with the opiate receptors in the brain to reduce your perception of pain and act similarly to narcotic-analgesic drugs such as morphine and codeine. *In contrast to opiate drugs, however, activation of opiate receptors by your body's endorphins does not lead to addiction or dependence.* There are more then twenty types of endorphins, some being up to thirty-three times more potent than morphine.[213] While enkephalins are a subfraction of the larger endorphin molecule, dynorphins (from the Greek dynamis = power) are also opioid peptides that are six to ten times more potent than morphine.[214] Dynorphins create feelings of euphoria as well as modulate how your body deals with stress. *Auricular therapy has been found to raise blood serum and cerebrospinal fluid levels of endorphins and enkephalins*[215] *and it has a proven neurophysiologic*

---

213 H H Loh, L F Tseng, E Wei, C H Li, 1976.

214 Han JS, Xie CW, 1984.

215 Pomeranz and Chiu, 1076; Sjolund it al., 1977, 1996; Clemente-Jones et al., 1979, 1980; Ng et al., 1975, 1981; Ho et al., 1978; Ho and Wen, 1989; Chen, 1993; Gao et al., 2008, 2011, 2012.

*and neurochemical basis.*[216] Dental pain thresholds were significantly increased by auricular electrical stimulation using the Stim Flex 400A transcutaneous electrical stimulation unit at appropriate auricular points.[217] Pain thresholds were not altered by sham stimulation at inappropriate auricular points. *In addition to relieving pain, endorphins enhance the immune system, reduce stress, postpone aging, modulate appetite, and lower blood pressure.* It is also fascinating to know that sound and music reduces complications after heart attack, calms anxiety, slows breathing, and increases production of endorphins[218] and 80 percent of stimuli that reach your brain comes in through your ears.

In 1973, H.L. Wen[219] discovered that auricular therapy facilitates the withdrawal from narcotic drugs, which led to numerous studies demonstrating the clinical use of this technique for substance abuse.[220] These findings led to the development of the National Acupuncture Detoxification Association (NADA) protocol using five ear points for the treatments of substance abuse. While auricular therapy has been shown to raise levels of met-enkephalin in human narcotic addicts,[221] Dr. Oleson[222] showed that while the occupation of opiate receptors sites by narcotic drugs lead to the inhibition of the activity of natural endorphins, auricular therapy facilitates the withdrawal from these drugs by activating the release of previously suppressed endorphins and in time, the opiate receptor sites return to their normal state. Professionally, I have had tremendous success with facilitating the withdrawal of patients from a multitude of prescription pain killers as well as with tobacco cessation using the NADA protocol along with some additional therapeutic points of treatment.

---

[216] Mayer et al., 1977.

[217] Simmons and Oleson, 1993.

[218] Gaynor M L, 1999.

[219] Wen H L, Cheung S, 1973.

[220] Dale R, 1993; Smith M, 1988.

[221] Clement-Jones V, McLoughlin L, Lowery P, Besser C, Rees L, Wen H, 1979.

[222] Kroening R, Oleson T, 1985.

Chapter 31

# It Can go Either Way

Unique to acupuncture, and particularly to auricular therapy, is that stimulating an active point on the ear can only make things better and will not exaggerate or worsen an existing condition. But how can this be? We know that conventional nerve stimulation or medications usually result in a unidirectional effect. For example, parasympathetic vagal stimulation only slows down heart rate and breathing while opioids only reduce digestive tract motility. However, acupuncture at PC-6 (pericardium point on the wrist) accelerates bradycardia and decelerates tachycardia. Acupuncture at ST-36 (stomach point on lower anterior leg) suppresses hyperfuntion (as in diarrhea) and stimulates hypofunction (as in constipation) of the gut motility.[223] Fortunately, *auricular stimulation is more sophisticated than a one-way, unidirectional effect because your brain and body are making the infinite number of decisions on how to best improve a condition and return it to a balanced, healthy state.* As I mentioned before, you have within you all that is essential to feel better. Your body just needs to be listened to and given the remedy that it's asking for. A critical component of successful treatment is the skill of the practitioner who must locate the precise acupoint(s) to treat as well as apply the appropriate frequency, duration, and amplitude. Due to the rich innervations of the outer ear, auricular therapy is perfectly suited to do this, since it has direct and indirect connections to the entire nervous system and all of the body systems.

Based on the morphogenetic singularity theory,[224] you can understand how auricular therapy regulates and if needed corrects bidirectionally other processes in the body by transmitting information from the stimulated ear points through meridians (electromagnetic energy channels) to the internal organs and to all other systems. The basic idea is that ear points, and all body points, and meridians are remnants of the growth control system, the first physical communication system in an embryo. The acupoints and meridians were intimately linked together during development, and they continue to stay connected and communicate. This growth control system directs the embryonic development, next to genetic imprinting. As every cell has its own place and function in the growing fetus, communication between cells is essential. Experiments with cells growing between a positive and negative electrode demonstrated that cells mainly grow in the direction of the negative electrode.[225] Cells can communicate with each other via gap junctions.[226] Increasing distance between cells impedes communication resulting in the breaking out of new groups of cells to

---

[223] Li, 1992.

[224] Shang, 1989, Shang, 2001.

[225] McCaig, 1987.

[226] Levin, 2007.

stay closer to one another. These groups are called organizing centers and determine the differentiation of other cells. The organizing centers have more gap junctions (channels extending across a gap between cells), a lower resistance to the flow of electricity, more superficial location on the embryo, and a more negative charge compared to the other cells. Collagen fibers can transport[227] impulses and form the communicating network between organizing centers. Thus, auricular points likely originated from organizing centers and are found on similar places on your body; they are part of a continuum. Their network of connecting collagen fibers is reflected in the meridian system, which has been proven to be present.[228] They can become visible by injecting radioactive tracers into acupuncture points.[229] Infrared measurements were also used to visualize the meridian system.[230] Additionally, most organizing centers are at the extreme points of curvature on the body surface such as the locally most convex points or concave points.[231] And almost all of the extreme points of the body surface curvature are acupuncture points. Thus, electrically active points (acupoints) on the skin are organizing centers revealed and organizing centers are acupoints concealed. They both share areas of high electrical conductivity, current density, high density of gap junctions, higher temperature,[232] higher metabolic rate, and carbon dioxide release.[233] Although there are several ways to stimulate an electrically active point on the skin, it is only bielectrode transdermal stimulation that will provide either a positive square wave (anode) or a negative square wave (cathode) to precisely treat the most active point(s). This will not only relieve pain, but it will also regulate it bilaterally (it can go either way) and return you to a state of health and balance. Incredibly, your auricle offers access to the remnants of an embryonic growth control system consisting of a network of growth control centers that communicate by way of an

---

227 Langevin, 2001; Langevin, 2002.

228 Ahn, 2005.

229 Vernejoul, 1984; Meng, 1989; Kovacs, 1992.

230 Liu, 1988.

231 Goodwin, 1969; Shang, 1989.

232 Comunetti A, Laage S, Schiessl N, Kistler A, 1995.

233 Bergsman O, Wooley-Hart A, 1973.

interconnected collagen network. This network remains today and is referred to as a meridian system. By stimulating an acupoint, the network of organizing centers throughout your body is activated and seeks to restore normal physiologic balance and function. This is a self-regulating process thus minimizes any risk of an "overshoot effect" of the therapy.

# Chapter 32

## Things Appear Simple from a Distance

In 1916, Albert Einstein went against the conceptions then in fashion when he proposed that light traveled in tiny bundles of energy called photons and behaved as a wave. He also stated that the shortest distance between two points could be a curved line. He once said, "In our endeavor to understand reality we are somewhat like a man trying to understand the mechanism of a closed watch. He sees the face and the moving hands, even hears it ticking, but he has no way of opening the case." In not so many words, this is at times, how I approach auricular medicine because there are some things that we cannot yet measure. The theory of relativity warns us against taking appearances, which hold only for a particular reference system, as the truth in any absolute sense. We should still recall how people reacted to the fact that the earth is spherical and later that the earth is moving around the sun. Certainly our perception did not agree with these facts.

Auricular medicine shares with the theory of relativity the importance of light as it travels as a wave in the form of a photon. Active electrical points of the ear emit bioelectromagnetic radiation in the form of measurable biophotons, or particles of light. The precise detection and treatment of the most active points requires the use of a bipolar electrode at specific frequencies and current strengths (millionths of amps) and either positive or negative square waves are used for treatment. Magnets are often used to encourage the generation of a microcurrent when gold or stainless steel semipermanent acupuncture needles are used for continuous therapy that lasts for several days. And since light is a wave (like a wave in the ocean) of alternating electric and magnetic fields, auricular therapy taps into both of these inseparable qualities to stimulate your mind and body to make changes to best improve a condition and return you to a balanced, healthy state.

It is often difficult both for doctors and for patients to understand and accept the concept that your entire body and mind can be represented and treated from a small remote area such as the external ear. And that the shortest distance between the pain in your lower back or foot and pain relief may be from the starting point of your ear, and not directly on your back or foot. Dr. Oleson[234] showed that relief of pain by auricular therapy can be best understood in terms of the stimulation-produced analgesia.[235] In addition to an ascending pain sensation pathway, there is also a descending pain inhibitory pathway that travels from the brainstem down the spinal cord and activates "suppressor cells" in the dorsal horn of the spinal column. By suppressing the pain messages at the spinal cord level,

[234] Oleson T., 1996.

[235] Liebeskind J, Mayer D, Akil H, 1995.

the messages are never forwarded to the higher levels of your brain for interpretation. *Thus, auricular therapy stimulates areas of the nervous system that inhibit pain messages from being processed and interpreted as painful.* This is known as the gate theory of pain[236] and like Albert Einstein's theory of relativity, it was met with considerable skepticism. Yet a half-century later, its basic principles remain unchanged.[237] Gate control theory explains how a stimulus that activates only nonnociceptive nerves can inhibit pain. The pain seems to be lessened when the area is rubbed because the activation of nonnociceptive fibers inhibits the firing of nociceptive ones in the laminae,[238] while in transcutaneous electrical nerve stimulation (TENS), nonnociceptive fibers are selectively stimulated with electrodes in order to produce this effect and thereby lessen pain.[239] *In auricular electrodermal stimulation, the peripheral nervous system, the central nervous system, and the electromagnetic energy system of meridians (that intimately communicate with every system in your body) are all engaged in the myriad of processes necessary to both inhibit pain and return your body to a balanced, healthy condition.*

---

[236] Metzack R, Wall P, 1965.

[237] Moayedi M, Davis K. D., 2013; Meldrum, Marcia L., 2015; Skevington S., 1995; Craig A.D., 2003; Cervero F., 2013.

[238] Kandel E R., Schwartz J H., Jessel T M., 2000.

[239] Ibid.

# Chapter 33

# Embryological Derivatives on the Ear

Our ear is a microcosm of our entire body. For each body part, there's a corresponding point on our ear. The main three nerves that innervate the ear also support the image of an embryological derivative: the endoderm, mesoderm, and ectoderm. The inner endodermal tissues of an embryo become the internal organs of the body. Disturbances in internal organs create an obstacle to the success of many medical treatments; thus these metabolic disorders must be corrected before complete healing can occur. Most of these deep embryological tissues are represented in the central valley of your ear, known as the central concha and are innervated primarily by the vagus nerve that serves mostly without conscious effort to regulate pain and pathology originating from your internal organs such as your heart, digestive, and respiratory systems. The middle mesodermal tissues of the embryo become the skeletal muscles, cardiac muscles, smooth muscles, connective tissue, joints, and bones. Nerves that process muscle pain, backaches, and headaches innervate this area, which is working both consciously and unconsciously to optimally react to the daily stressors of life. Musculoskeletal pains often become chronic and can refer additional pain throughout your entire body. The outer ectodermal tissues of the embryo become the outer skin, endocrine glands, and our nervous system. This embryologic tissue is represented on the outermost area of your ear and includes the outer rim of your ear and the ear lobe. These tissues are impacted by both physical challenges and psychological reactions and regulate neuropathic pain such as peripheral neuropathy and psychological pain such as depression.

It's important to realize that the upside down representation of our body on the ear is only present during *Nogier Phase 1*, which is used to treat the majority of medical conditions and somatic tissue disorganization to include *acute pain*. *Nogier Phase II* shows the upright man pattern (as opposed to the upside down man in Phase I) and is used to treat more difficult, *chronic conditions* that have not responded successfully to treatment of the Phase I microsystem points. This phase is useful for correcting CNS dysfunctions and mental confusion that contributes to the psychosomatic aspects of pain and pathology of chronic illnesses. *Phase III*, the horizontal man pattern, is used the least frequently, but it can be very effective for relieving unusual conditions. Phase III points often indicate *prolonged inflammatory conditions* that become *degenerative* and affects basic cellular energy. Nogier's three phases account for some of the discrepancies between the French and the Chinese ear acupuncture charts. *There is a Phase IV, which refers to the posterior of the ear and affects all of the muscles of our body.* This phase has the body oriented upside down just like in Phase 1. In my experience, it's always best to treat the back of the ear first to address any muscular component of pain prior to treating the front of the ear, which addresses any sensory component to pain.

Paul Nogier, M.D., proposed the phase theory in the early 1980's. When any point on the ear is treated, one primary effect and two secondary effects are often observed. If the primary effect of treating an area is endodermic (internal organs), the secondary effects will be mesodermic (musculoskeletal) and ectodermic (nervous system, endocrine system and the skin). If the primary effect of treatment is mesodermal, the secondary effects will be ectodermal and endodermal. If the primary effects of treatment are ectodermal, then the secondary effects will be endodermal and mesodermal. Thus, treating a point in a particular region of the ear will also affect the other two regions.

Chapter 34

# Two-Way Reflexes on the Ear

In the 1970s, research described the external ear as a *microacupuncture system*[240] that connects peripheral regions of the body to the central nervous system. Dr. R. Dale described the ear as exhibiting both *organocutaneous reflexes*, which allow the outer ear to reveal underlying body pathology (visual changes, tenderness, electrodermal differentials), and *cutaneo-organic reflexes, which enable stimulation of the outer ear to heal the pathology or imbalance.* So your ear is like a peripheral microcosm of your entire body that connects to the CNS. The CNS works to institute corrective measures intended to normalize any disordered neural circuits, but strong environmental stressors or intense emotions may cause the central nervous system circuitry to break down. Tsun-nin Lee[241] showed that pathological changes in peripheral tissue may lead to dysfunctional nerve firing patterns in the corresponding neural circuits in the brain and spinal cord. *If these neural patterns or reflexes are impaired, this becomes the new normal and peripheral disease may remain chronic.* Chronic pain and disease are likely learned, maladaptive programming of these neural circuits. Stimulation of the outer ear serves to induce reorganization of these pathological brain pathways to return your body to a balanced, healthy state and to heal the pathology.

A double-blind study of auricular points that correspond to heart disease showed a significantly higher frequency of reactive ear points at the Chinese heart points in the inferior endodermal concha

240 Dale R, 1976.

241 Lee TN, 1977; Lee TN, 1994.

region and on the tragus for patients with myocardial infarctions and angina pain than for a control group of healthy subjects.[242] There was no difference between the coronary heart disease group and the control group in the electrical reactivity of auricular points that did not represent the heart. A study by Dr. Oleson[243] also supported the concept that Dr. Nogier's mesodermal antihelix zone on the ear corresponds to musculoskeletal pain. There is also a compelling connection between auricular therapy and the area of the brain classically associated with the neurophysiologic regulation of weight control, the hypothalamus. The two regions of the hypothalamus that regulate weight are the ventromedial hypothalamus, which provides a sense of fullness or hunger, and the lateral hypothalamus, which when stimulated elicits you to start eating. Inhibition of vagus nerve activity as well as stimulating serotonin release by stimulating the ear at the appropriate sites, likely delays the stomach from emptying (measured by a reduction in peristaltic waves)[244] leading to a sense of fullness and early satiety or satisfaction. Sun and Xu[245] demonstrated that the reduction in body weight by the auricular acupuncture group was significantly greater than shown by the control group, while the triglyceride blood lipid levels were also diminished in the auricular therapy group. Richards and Marley[246] also found that weight loss was significantly greater for women in an auricular therapy group than in a control group.

In summary, an organo-cutaneous reflex occurs when there is pathology in the body, which sends neurological messages to the central nervous system, and these messages are then sent to a point on the skin of the ear. A cutaneo-organic reflex works in the opposite direction. Here, when a point on the skin of the ear is stimulated, neurological messages to the central nervous system are activated, which send messages to a particular area of the body. Thus the ear provides a two-way reflex.

---

242 Saku K, Mukaino Y, Ying H, Arakwa K, 1993.

243 Oleson T, Kroening R, Bresler D, 1980.

244 Choy D S, Eidenschenk E, 1998.

245 Sun G, Xu Y, 1993.

246 Richards D, Marley J, 1998.

# Chapter 35

## Diagnostic Value of Visual Changes on the Ear

Practitioners of auricular therapy emphasize the diagnostic value of visually examining the outer ear.[247] Shiny spots of different colors on the ear surface typically indicate acute inflammations in your body. If pressure is applied to these colored areas, they are usually painful. The absence of these spots does not indicate the absence of any medical problem, but the occurrence of colored regions on the ear does signal a high probability of some pathology in the correspondent part of the body. Dandruff-like areas or peeling skin on the outer ear often indicate a chronic condition in your body and if the treatment is effective for healing that condition, the flaky regions on the ear do not reappear. Diagonal folds in the skin over the ear lobe have been correlated with coronary problems. In double-blind clinical studies published in the *New England Journal of Medicine*, Lichstein et al. (1974) and Mehta and Homby (1974) correlated diagonal ear lobe creases with coronary problems. Their studies showed that *the presence of a crease running diagonally on the earlobe was more predictive of the occurrence of a coronary problem then knowing either the patient's blood pressure level or serum cholesterol level.* The presence of a diagonal earlobe crease (DELC) has been proven to be of prognostic significance in more than forty separate medical studies.[248] Earlobe creases have historical significance as they can

---

247 Kvirchishvili, 1074: Romoli and Vettoni, 1982; Huang, 1996.

248 Friedlander A H, Lopez-Lopez J, Velasco-Ortega Eugenio, 2012.

be seen in the highly accurate and detailed Roman sculptures of emperors. The sculpture of Roman Emperor Hadrian prominently displayed bilateral earlobe creases,[249] which support classical writings that suggest he died from congestive heart failure resulting from hypertension and coronary atherosclerosis. Sculptures of other Roman emperors and nobles in museums do not have diagonal earlobe creases. The majority of clinical and angiography reports support the premise that the diagonal earlobe crease is a valuable extravascular physical sign able to distinguish patients at risk of succumbing to atherosclerosis of the coronary arteries. More recently, reports using B mode ultrasound have also linked the DELC to atherosclerosis of the carotid artery. In the *New England Journal of Medicine*,[250] earlobe crease and coronary atherosclerosis had a 72 percent correlation. Furthermore, the prevalence of diagonal earlobe crease increased with severity of coronary artery atherosclerosis (in 79 percent of patients with double or triple vessel disease). Relative to specific known atherogenic risk factors, diagonal earlobe crease has also been reported to be associated with hypertension[251], high serum cholesterol[252], smoking, obesity and diabetes mellitus[253].

In 2010, a dental researcher reported on the prevalence of diagonal earlobe crease in neurologically asymptomatic patients and the presence of calcified carotid fatty deposits on their dental radiographs.[254] This demonstrated that it would be prudent for dentists to observe their patients ears for the presence of diagonal earlobe crease and in conjunction with their medical history, vital signs and panoramic radiograph, formulate a risk assessment and determine if medical consultation is indicated for evaluation of coronary and/or carotid artery disease. The patients of mine that

---

249 Petrakis NL, 1980.

250 Kaukola A, Manninen V, Malle M, Halonen Pl, 1979.

251 Toyosaki N, Tsuchiya M, Hashimoto T, Kawasaki K, Shiina A, Toyooka T, 1986; Merlob P, Amir J, Reisner SH, 1981; Moncada B, Ruiz JM, Rodriguez E, Leiva JL, 1979; Kristensen BO, 1980

252 Frank ST, 1973; Toyosaki N, Tsuchiya M, Hashimoto T, Kawasaki K, Shiina A, Toyooka T, 1986

253 Frank ST, 1973, Doering C, Ruhsenberger C, Phillips DS, 1977; Raman R, Rani PK, Kulothungan V, Sharma T, 2009

254 Friedlander AH, 2010.

have a prominent earlobe crease are often either being treated for cardiovascular disease or have a family history of coronary problems. A few times, patients were unaware of being at risk but after noting the presence of an earlobe crease along with their particular vital signs, I urged them to seek further attention. This resulted in a more thorough evaluation and ultimately, many of them were diagnosed with coronary problems and treated appropriately.

In summary, since a comprehensive review of modern research has demonstrated the mechanisms of auricular medicine and confirmed its efficacy in a multitude of disorders, it no longer seems possible to doubt its truth. Your brilliant and intricate ear does far more than simply serve as a wave catcher. It provides unprecedented access to your nervous system and immune system to help alleviate pain and revive and accelerate the healing process naturally.

Chapter 36

# Illustrations

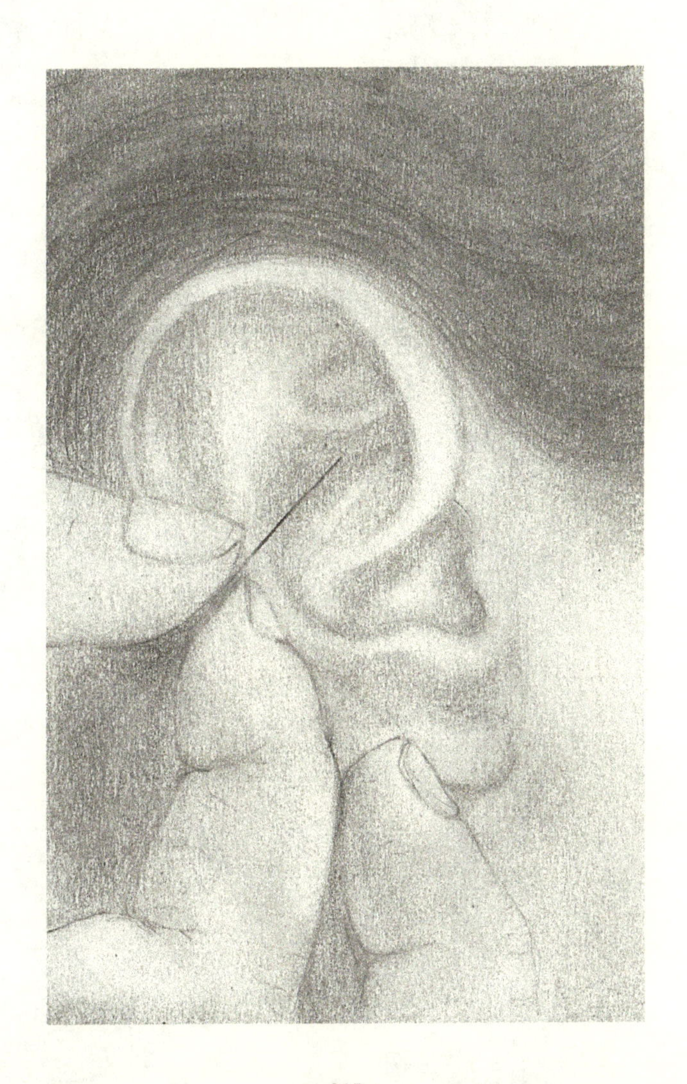

L.C. Halfus

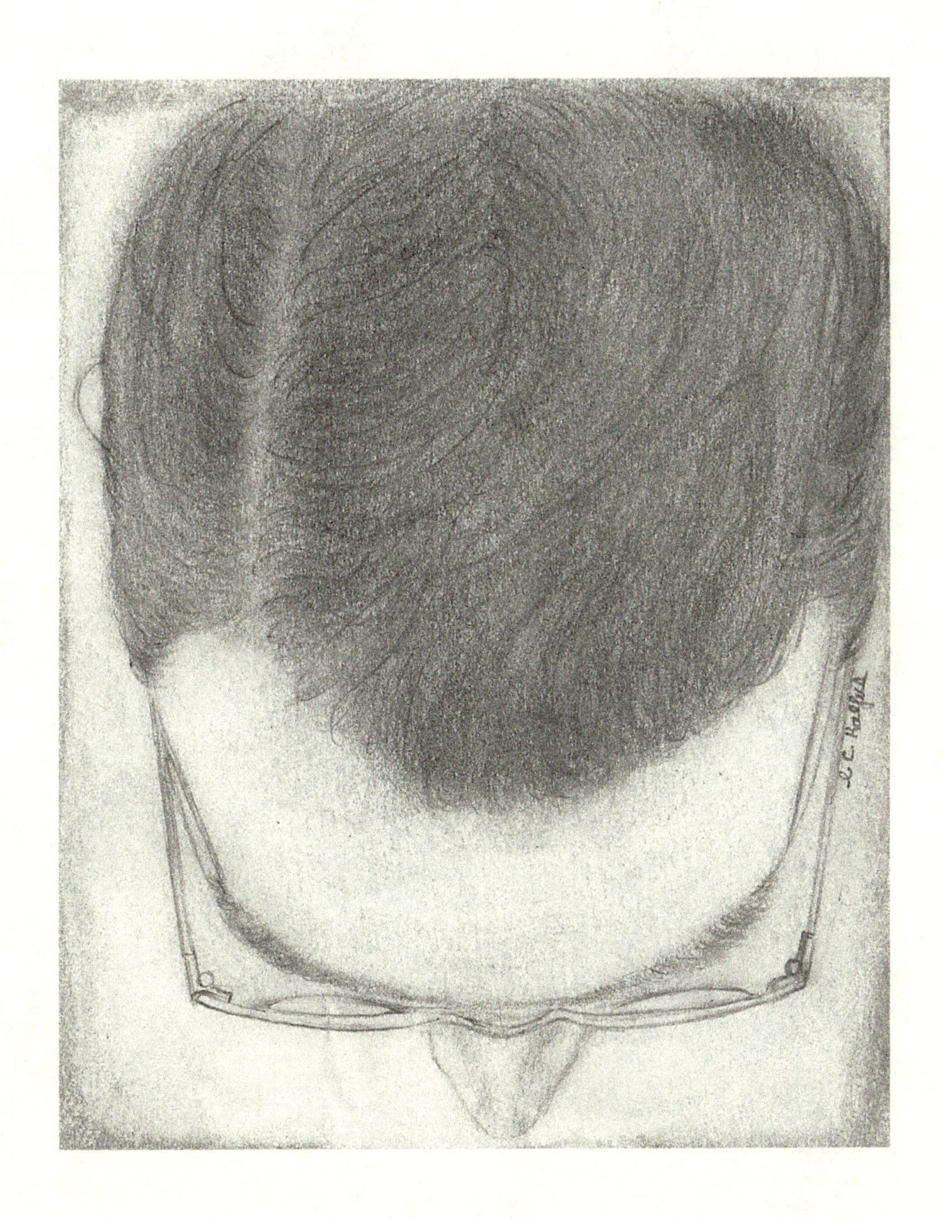

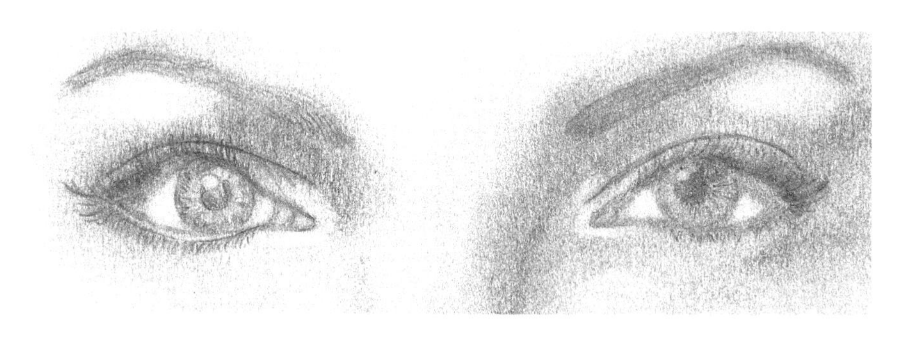

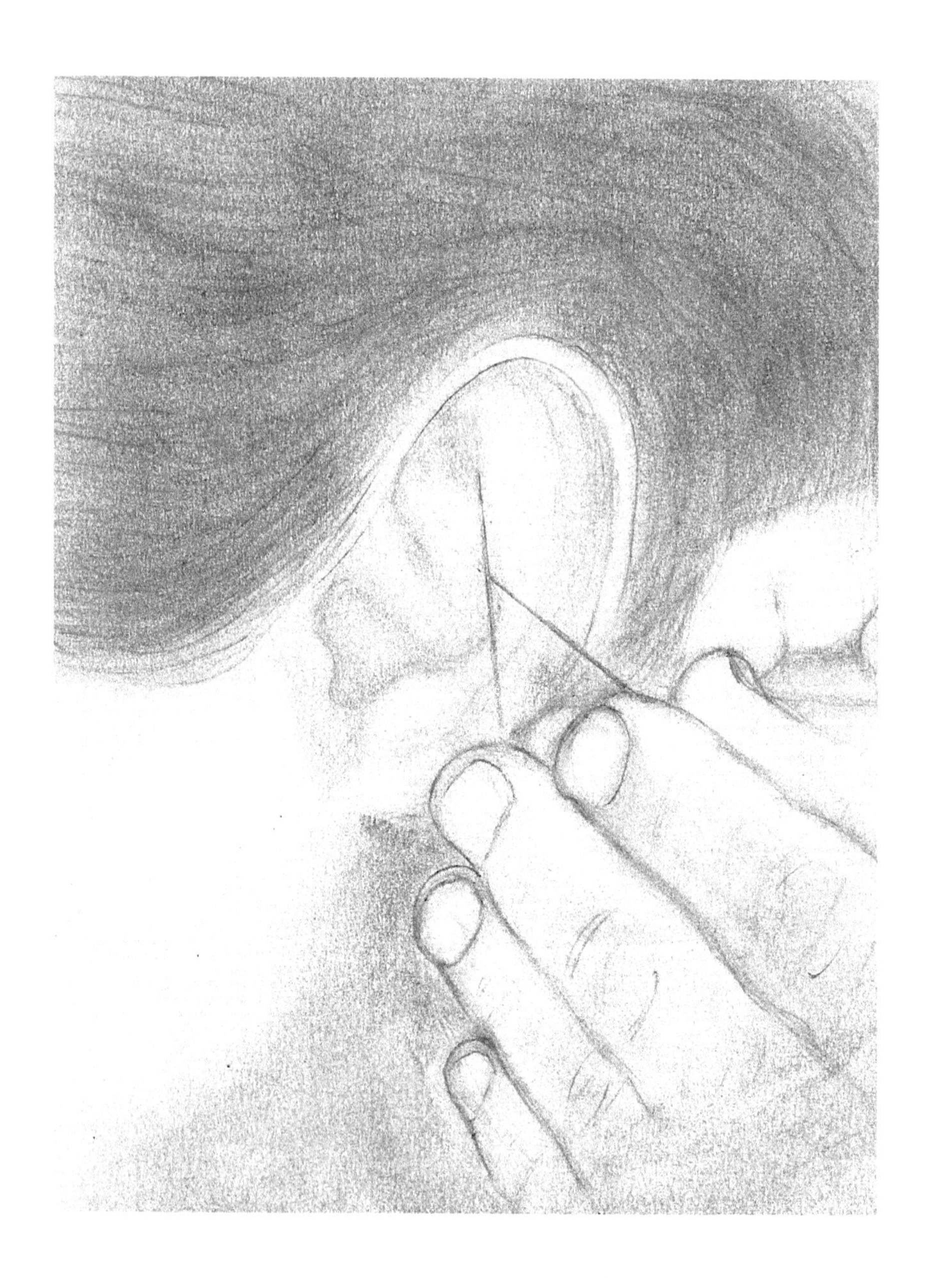

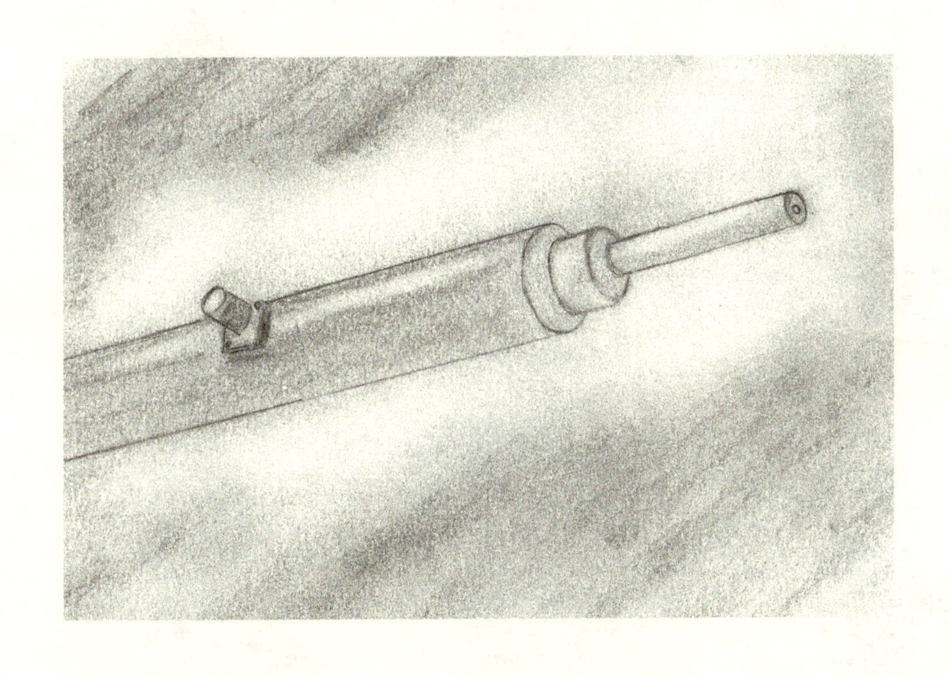

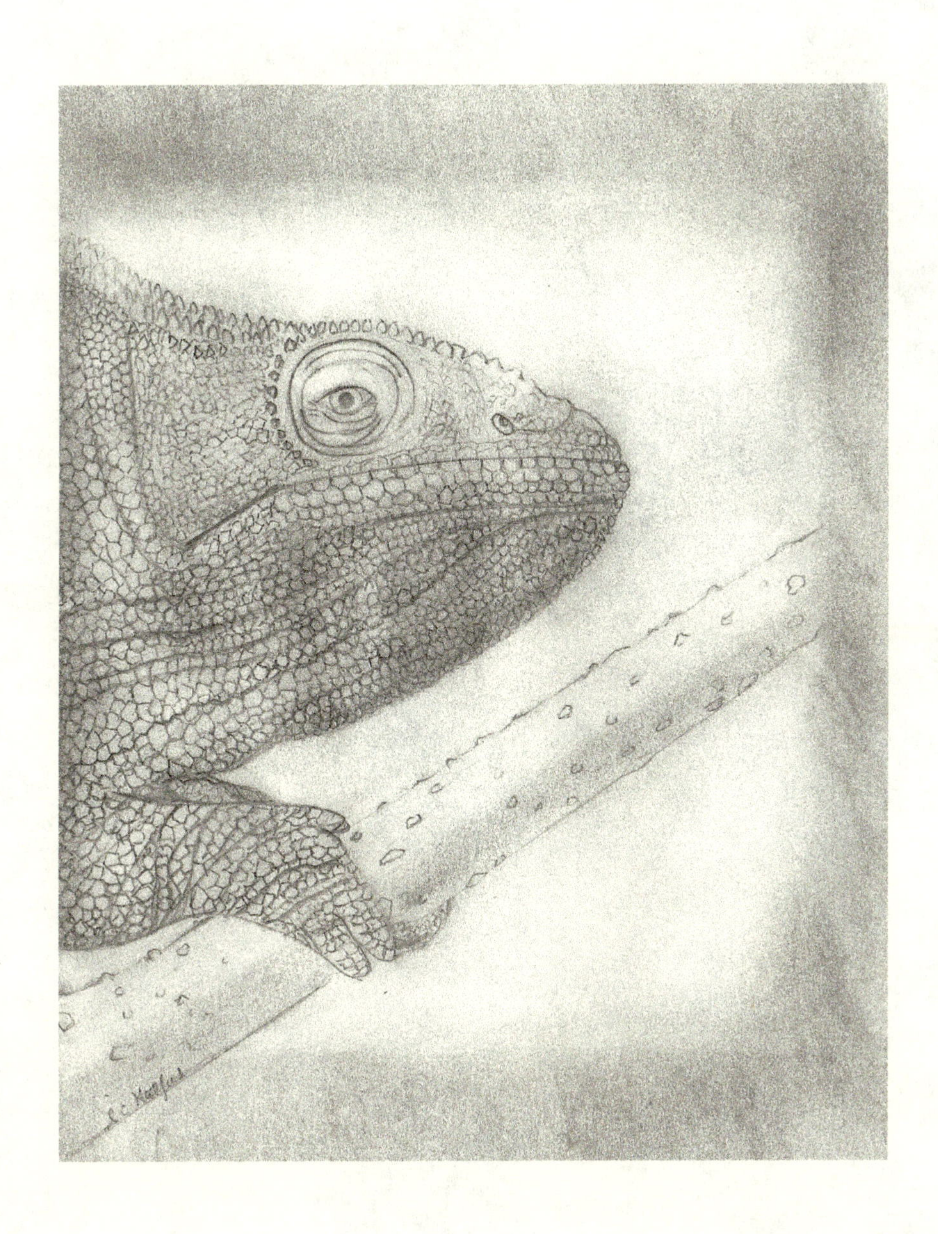

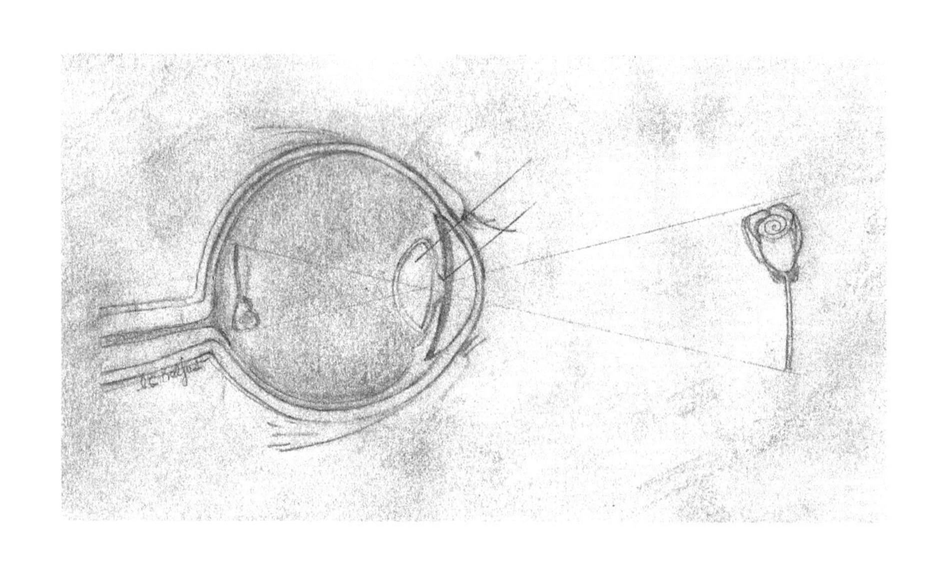

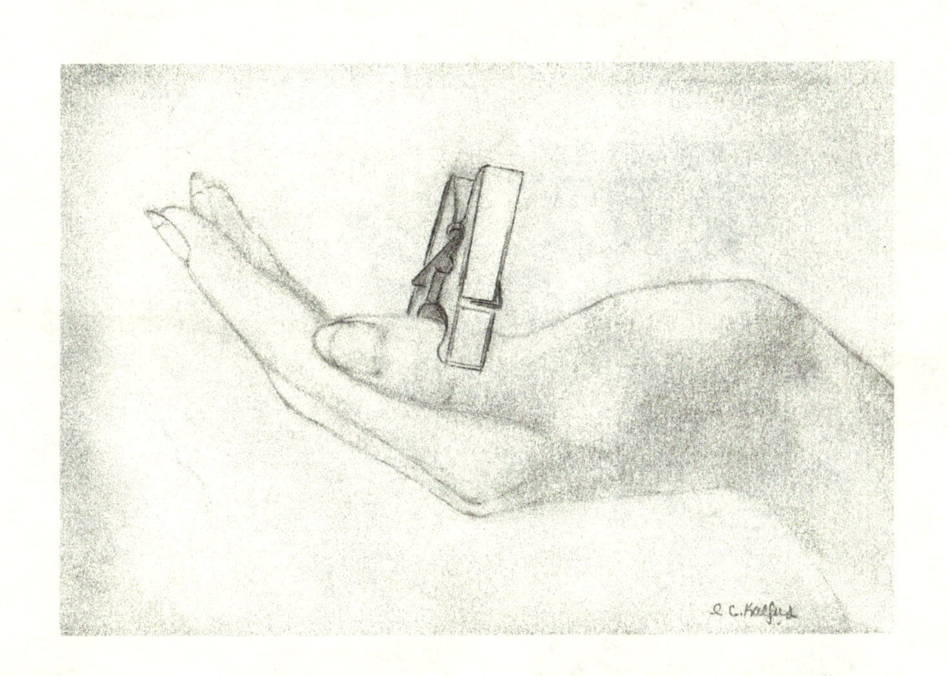

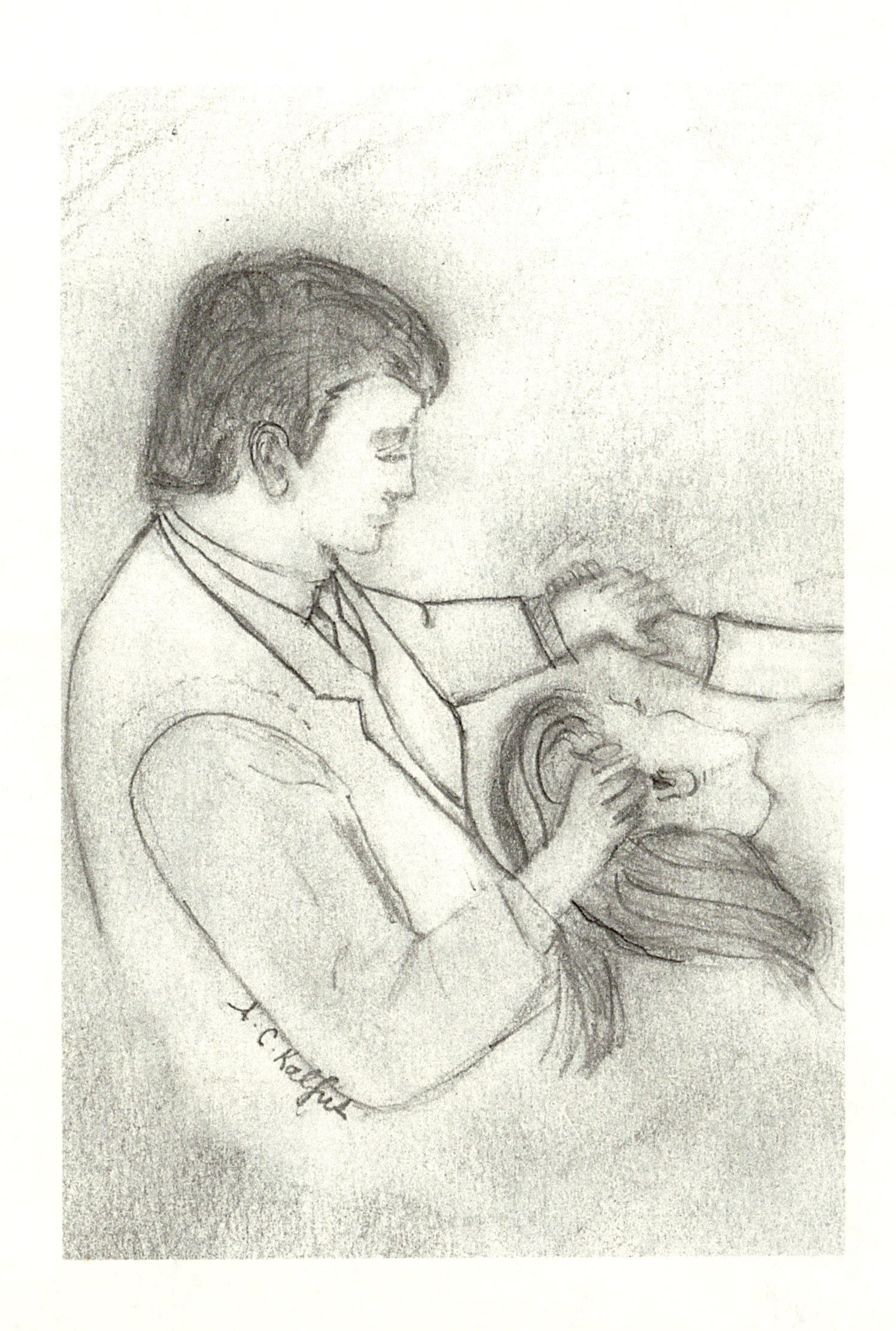

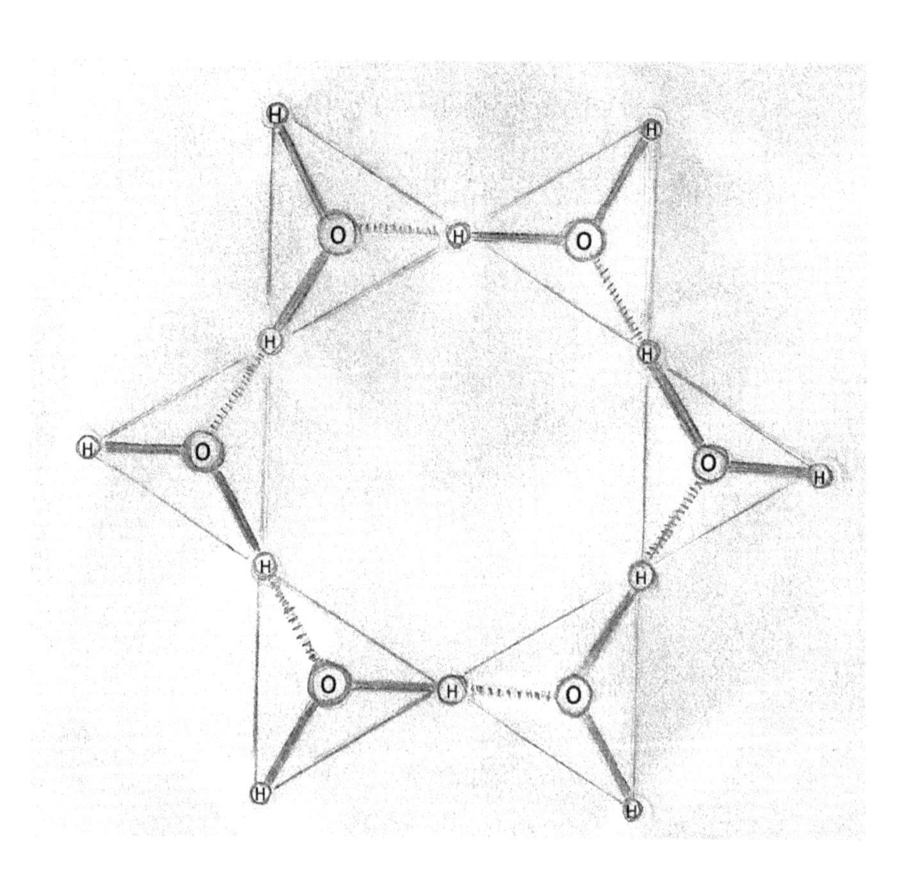
H
H
O
H
O
H
H
O
H
O
H
H
O
H
O
H
H

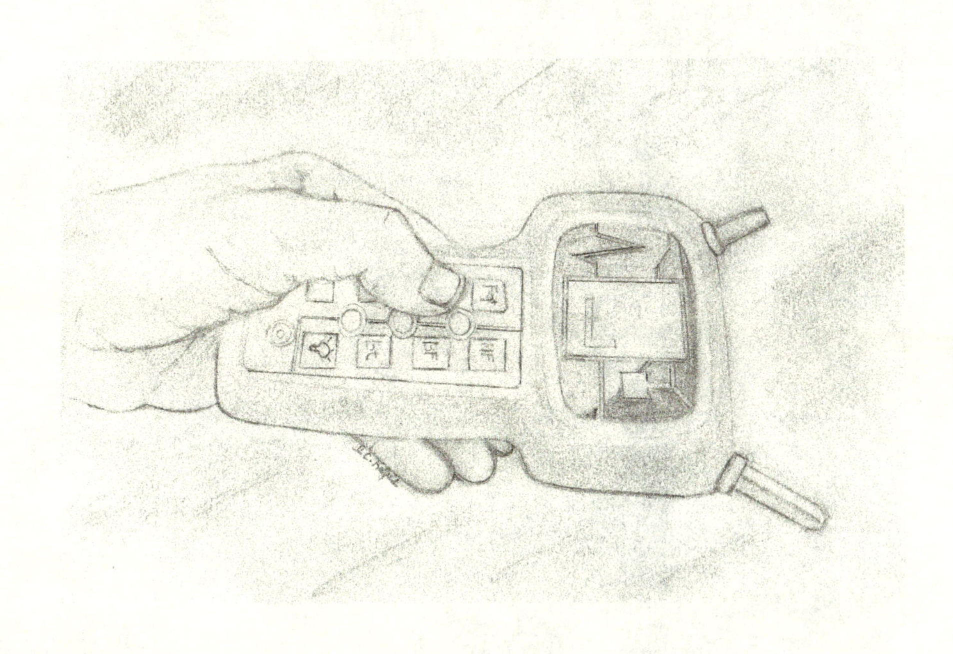

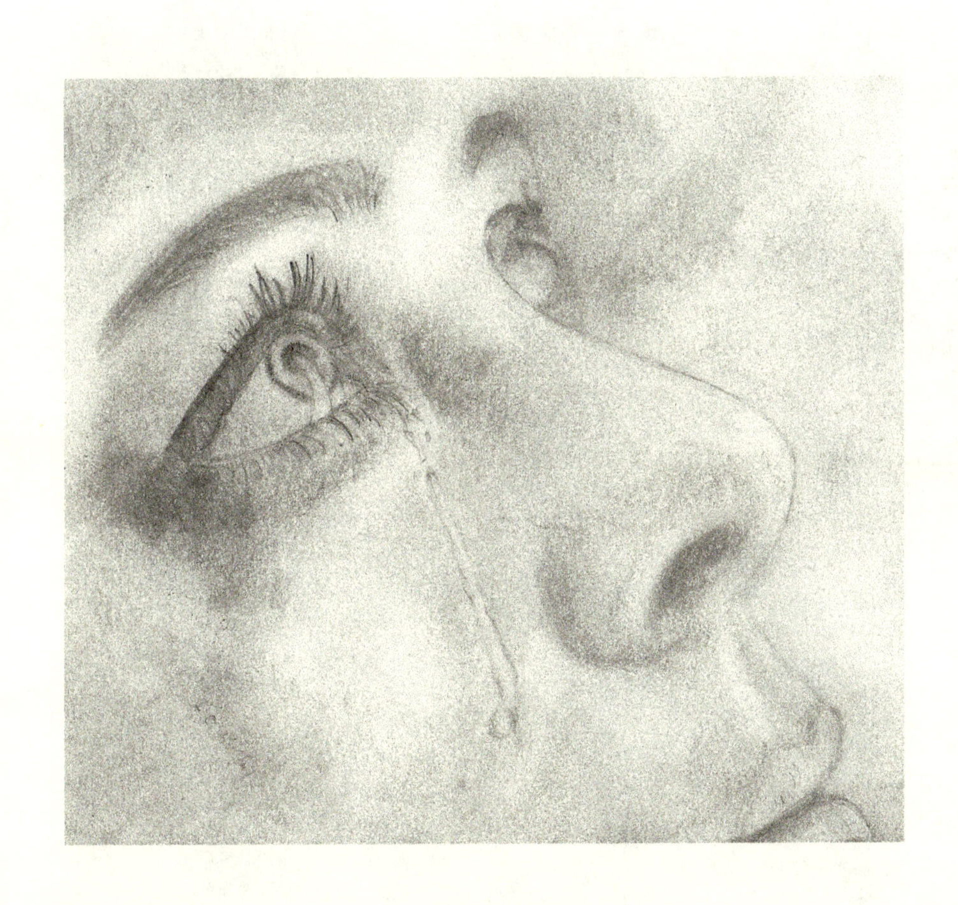

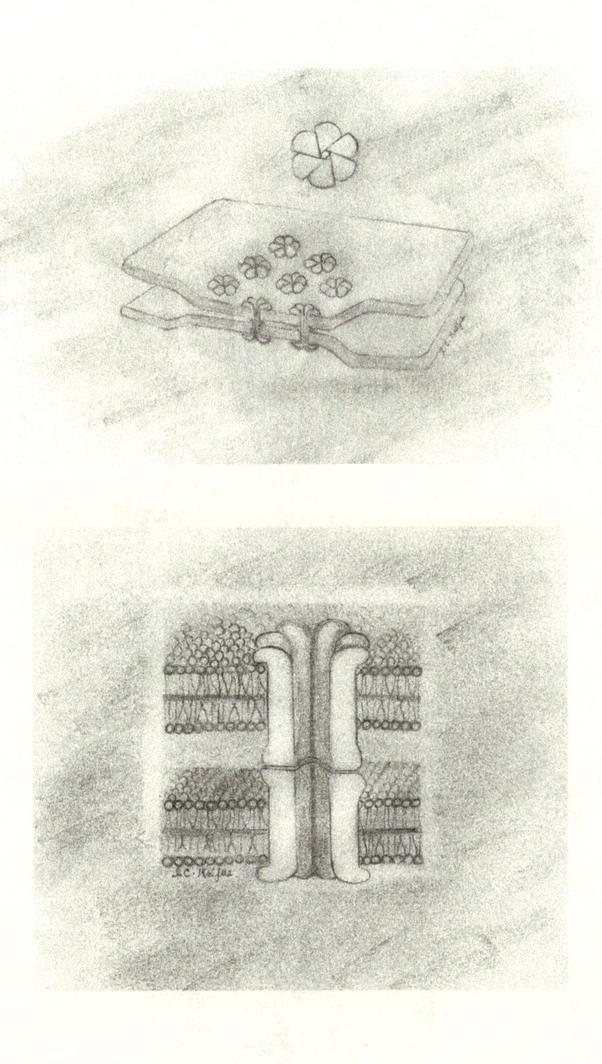

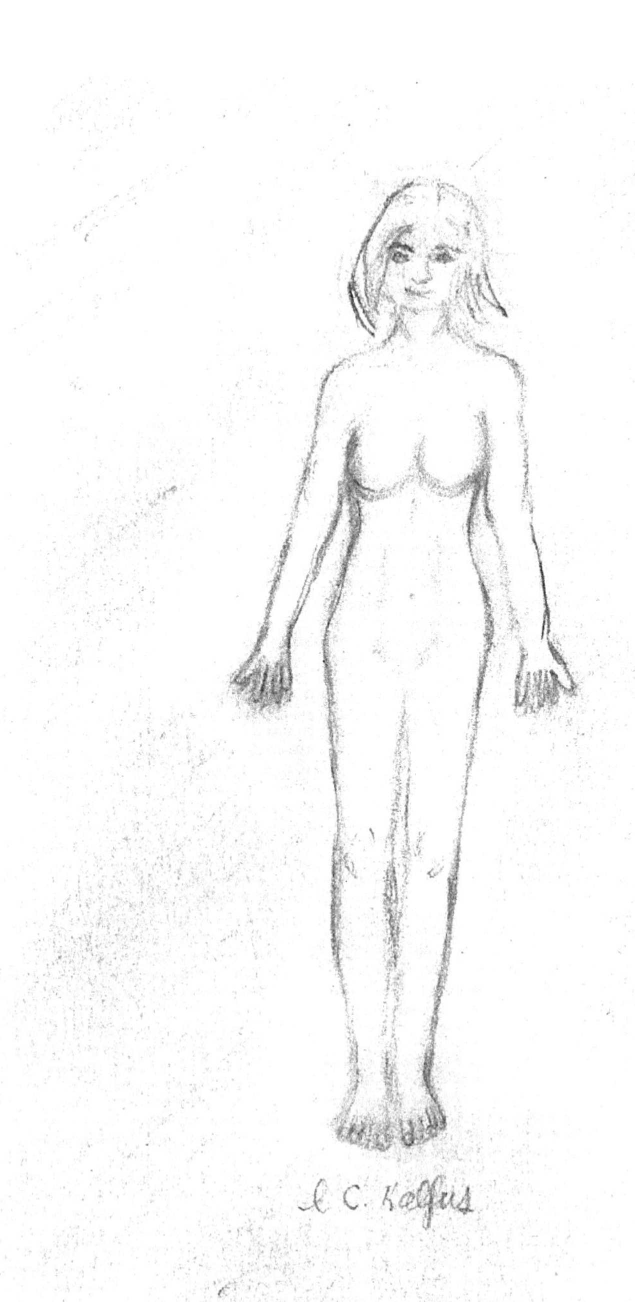

Thank
You

# Index


## A

Achilles heel pain, 123, 128
ACTH (Adrenocorticotropic Hormone), 154
acupuncture, xiv, 8, 43, 49–51, 53, 61–62, 67–69, 132, 136–37, 153–57, 160–62, 179–82, 185–95, 197–98, 201–3
acupuncture analgesia, 153–54, 182, 195, 198
adrenaline, 145–46
aging, 21–22, 164, 168, 178
allergies, 30–34, 141–42, 147
aluminium, 172, 178
Alzheimer's disease, 178
American Academy of Pain Medicine, 64
analgesia, 153–54, 161, 180, 182, 193, 195–96, 198, 206
anode, 194, 203
antibiotics, 26–27, 30
antioxidants, 164
arteries, 76, 87, 89, 94, 129, 141–42, 144, 146
arterioles, 141, 155, 183
ASP needle, 120–21
atherosclerosis, 213
atoms, 169–70, 176, 179
ATP (Adenosine Tri-Phosphate), 164
Aura Humanus Tractatus (Valsalva), 50
auricle, 180–81, 185–86, 195, 203
auricular medicine, x, xii, xiv, 2–5, 24, 40, 43, 46, 53, 56, 58, 67–69, 93, 98, 206
auriculotherapy, 53, 137, 154, 159, 161–62, 181, 192
autoimmune diseases, xii, 28–29, 34
autoimmunity, 28

## B

biophotons, 151, 159–60, 162–66, 186, 189, 206
bipolar electrodes, xiv, 71, 139, 180, 183, 191–92, 194, 206
blood pressure, 31, 33, 68, 133, 146
bloodstream, 28–31, 34, 145, 177
body temperature, 8, 19, 87–88
Bourdiol, Rene, 53, 181
brain activity, 12–13, 15, 54, 61–62, 196–97
brain communication disorders, 111
brain imaging, 195

## C


Caesar, Gaius Julius, 46, 53
calcium, 32, 171, 177–78
Carlson, Charles, 16
cauterization, 47, 50
celiac disease, 28
centrioles, 93
CER (cutaneous electrical resistance), 183
chronic pain, 4, 11, 57–58, 64–65, 71, 96, 98, 117–18, 123, 126, 128, 136, 151, 157–58, 182
Civil War, 50, 137
Clegg, James, 172
CNS (central nervous system), 154, 158, 195, 198, 207, 210–11
colors, 88, 97–98, 145, 165, 212
crystals, 170, 172


## D


dairy products, 30, 32
Dark Ages, 48
Darwin, Charles, 1
DC resistance, 185
dental focus, 118, 123–28
depression, ix, 32, 45, 58, 68, 109, 111–13, 115, 208
diaphragm, 15–16
diets, 25–27, 30, 32–34, 164
  avoidance, 33–34
digestive tract, 28–30, 144
DNA, 13, 23, 151, 164, 166, 169
  mitochondrial, 164
Drost-Hansen, Walter, 172
dynorphins, 154, 181, 199


## E


ear acupuncture, 51, 53, 137, 160, 180–81, 209
ear infections, 4–5, 33
earlobe creases, 212–14
ear pain, 79, 82
Ebers Papyrus, 44–45
ECG (electrocardiogram), 159
EEG (electroencephalogram), 159
Egypt, ancient, 43–45
Einstein, Albert, 205
electrical activity, 4, 60, 67, 75–76, 86, 88, 141–42, 160
electrical conductivity, 160, 183, 186, 188, 191
electrical stimulation, 74, 79, 89, 127–28, 136, 154, 190, 200
electricity, 74–76, 84, 87, 134
electrodes, 85, 185, 192, 207
electroencephalogram, 159
electromagnetic field, 134, 162, 166, 189
electrons, 164
embryo, 92, 149, 192, 202–3, 208
emotions, 6, 13–14, 57–58, 96, 98, 133
endorphins, 65, 154, 162, 180–81, 193, 199–200
energy, 51, 91, 93, 146, 151, 162, 164, 171, 176–77, 179, 205
enkephalins, 154, 180–81, 199
Erismann, Theodore, 38


## F


first rib syndrome, 118–23, 128
food allergies, 29–34, 121

food-borne illness, 25
food proteins, 29–31, 33–34
Franklin, Bache, 137
Franklin, Benjamin, 50, 137

**G**

ganglion, 118–19
gap junctions, 174, 176, 188, 192, 202–3

**H**

heart rate, 68, 133, 146
hexagonal water, 22, 168, 170
Hippocrates, 46
Hippocratic medicine, 46
hormones, 94–95, 133, 145–46, 165, 196, 198
hospitality, 48
Huang, Helean, 181
humanity, 32, 112, 133
hydrogen bonds, 169
hyperactivity, 111, 113, 115
hypnosis, 153
hypothalamus, 87, 95, 144, 161, 180, 211

**I**

IgE (Immunoglobulin E), 31–32
IgG (Immunoglobulin G), 31
immune cells, 19, 29, 31, 89, 146
Immunoglobulin E (IgE), 31–32
Immunoglobulin G (IgG), 31
immunoglobulins, 30–31
infrared light, 88, 129, 177

**L**

Langerhans cells, 89, 146
laterality disorders, 100, 109, 112, 115
Leriche, Rene, 142
light of life, 5, 92–93
light photons, 88, 90
light pulsing, 93–96, 111, 178
light therapy, 68, 92, 99, 115, 128, 145, 178
Ling, Gilbert, 172
liquid crystals, 170–73
lymphocytes, 29, 34

**M**

meridians, 51, 67, 184, 187–88, 193, 202–3, 207
microsystems, 131
midbrain, 180, 197
milk, 30, 32–33
mitochondria, 164
morphine, 199
morphogen gradients, 189
motility, 94, 201–2
musculoskeletal pains, 208, 211

**N**

NADA (National Acupuncture Detoxification Association), 200
Naloxone, 153–54
narcotic-analgesic drugs, 199
NCHS (National Center for Health Statistics), 27, 64
nerves, vagus, 68, 133, 208

neurophysiology, 151, 159
neurotransmitters, 154, 196, 198–99
neurovascular points, 74, 179, 183–84
Nixon, Richard, 53, 137
NMR (Nuclear Magnetic Resonance), 23, 168, 173
nodes of Ranvier, 160
Nogier, Paul, 2, 6, 8, 43, 50, 52–53, 82, 97, 123, 131, 137, 139–42, 162, 179, 181
Nogier, Raphael, x, xiv, 6, 77, 92, 100, 109, 111, 123, 137, 142, 145
Nogier pulse, 47, 129–30, 142–43

## O

Oleson, Terry, x, xiv, 53, 181, 192
opioids, 153, 201
osteoporosis, 32–33

## P

PAG (periaqueductal grey), 161–62, 195–96
pain
  nociceptive, 160
  sciatic, 43, 50, 52
pain management, ix, xiv, 2, 14, 19, 55, 57, 157
pain relief, 15–16, 50, 60, 136, 153–54, 162, 181, 198, 206
Parkinson's disease, 94
pathology, 39, 67, 84, 87, 98, 128, 191, 208–12
PET (positron emission tomography), 161–62, 197
pharmacy, 7, 44
photons, 88, 90, 146, 159, 163–64, 205–6
Pollack, Gerald, 171–72
proteins, 23, 27–32, 34, 164, 166, 171
public health, 27, 64
pulsed light, 3–4, 8, 71, 93, 100, 129, 136
pulse strength, 8, 47, 68, 71, 76–78, 97–101, 111–12, 127–28, 142

## R

rabbit test, 145–46
RAC (Rockefeller Archive Center), 51
red light, visible, 177
reflexes, 98, 141–42, 152, 161, 191, 210
reflex points, 73, 78–81, 83, 160, 185, 192
rib, first, 118–23
Rockefeller Foundation, 51
Roman Empire, 46–47, 53–54, 141

## S

sensitivity, 30, 32, 60, 76, 84, 98
Shores, Jim, xiv, 53, 139, 192
shoulder pain, ix, 64, 79, 120–21, 123–24, 127–28, 136
skin disorders, 32–33
somatotopic inversion, 37
SPECT (Single Photon Emission Computed Tomography), 197
spinal cord, 8, 71, 76, 79–81, 95, 118, 180, 196, 206, 210
stellate ganglion, 4, 118–19, 122–23

stimulation, 60, 67, 72, 151, 154–55, 180, 184, 194–95, 198, 210
Stratton, George, 37
sunlight, 6, 22, 93, 95, 117, 151, 168
sweating, 186
symmetry, 94–95, 98, 170
Szent-Gyorgyi, Albert, 172

**T**

teeth, 118, 123–24, 126–28
thalamus, 79, 161–62, 195, 197–98
T-helper cells, 28
Tinel, Jules, 146
toxins, 28, 89, 171–72

**U**

U.S. National Institutes of Health, 3

**V**

Valsalva, Antonio Maria, 50
VAS (Vascular Autonomic Signal), 47, 53, 141

**W**

water clusters, 23, 169, 173
water molecules, 22–23, 168–73, 176, 178–79
wavelengths, 165, 177
Wexu, Mario, 181
white light, 97, 129, 147, 177
World Health Organization, 3, 132, 136, 182, 190

Printed in the USA
CPSIA information can be obtained
at www.ICGtesting.com
LVHW041631181024
794185LV00022B/242/J